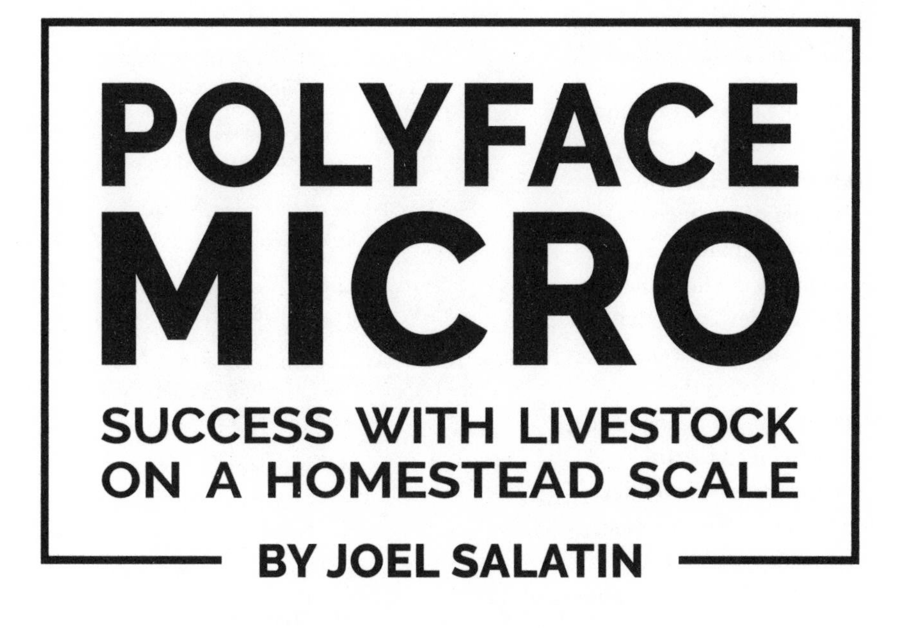

Foreword by Justin Rhodes

PHOTO: Rachel Salatin

JOEL SALATIN and his family operate Polyface Farm in Virginia's Shenandoah Valley. His parents purchased the farm in 1961 and developed the basic principles of design and production that now show 60 years' refinement. He gravitated toward communication activities in high school and college, graduating with a BA degree in English, and after a brief journalism hiatus returned to the family farm full time Sept. 24, 1982. Editor of *The Stockman Grass Farmer* magazine, he writes and speaks around the world on food and farm issues. With a long track record of innovation and excellence, Polyface holds educational seminars, farm tours, day camps and events to encourage duplication and understanding. This is Joel's 15th book.

Other Books & Resources by Joel Salatin

Polyface Designs: A Comprehensive Construction Guide for Scalable Farming Infrastructure

Beyond Labels: A Doctor and a Farmer Conquer Food Confusion One Bite at a Time

Your Successful Farm Business: Production, Profit, Pleasure

The Marvelous Pigness of Pigs: Nurturing and Caring for All God's Creation

Fields of Farmers: Interning, Mentoring, Partnering, Germinating

Patrick's Great Grass Adventure: With Greg the Grass Farmer

Folks, This Ain't Normal: A Farmer's Advice for Happier Hens, Healthier People, and a Better World

Holy Cows and Hog Heaven: The Food Buyer's Guide to Farm Friendly Food

You Can Farm: The Entrepreneur's Guide to Start and Succeed in A Farming Enterprise

Salad Bar Beef

The Sheer Ecstasy of Being a Lunatic Farmer

Pastured Poultry Profits: Net $25,000 on 20 Acres in 6 Months

Everything I Want to Do Is Illegal: War Stories from the Local Food Front

Family Friendly Farming: A Multi-Generational Home-Based Business Testament

The Polyface Farm DVD

Polyfaces: A World of Many Choices DVD

Primer Series: Pigs 'n Glens DVD

Primer Series: Techno Stealth: Metropolitan Buying Clubs DVD

The Salatin Semester: A Complete Homestudy Course in Polyface-Style Diversified Farming

POLYFACE MICRO:

SUCCESS WITH LIVESTOCK ON A HOMESTEAD SCALE

By Joel Salatin

Foreword by Justin Rhodes

Polyface, Inc.
Swoope, Virginia

Polyface Micro: Success with Livestock on a Homestead Scale

First Edition, 2021

ISBN: 978-1-7336866-2-4
Library of Congress Control Number: 2021911577

Edited by Jennifer Dehoff.
Cover artwork by Rachel Salatin.
Cover design & book desktop publishing by Jennifer Dehoff Design.

Printed and bound in the USA.

CONTENTS

Foreword

By Justin Rhodes

I was in the airport on my way to a speaking engagement when the coronavirus got my attention for the first time. How could I miss it? It was on every TV, in every room. China was in a major lockdown. Before we knew it, that news had spread to America where gatherings were limited, quarantines in place, food shortages widespread, mask mandates enforced, and now, talks of vaccine passports.

I've been teaching folks how to homestead for years and the usual appeal has been the ability to save money, connect with their food source, get healthy or do something positive together as a family. However, the case for growing food became much more urgent in 2020 when a phrase like "food shortages" entered our vernacular for the first time. Folks were rushing towards the homestead movement for a new, very powerful reason: food security. Seed companies were backed up, homestead supplies stores out of stock and hatcheries were filling every nook and cranny with incubated eggs. It's like homesteading went from a novelty on the outskirts of society to an awesome trend that included real solutions to some of our newfound problems.

No doubt the entire world has gone over some sort of

precipice. And it's likely you've gone through your own individual struggle, whether it's related or not. My permaculture teacher, Geoff Lawton pointed out that it usually takes going over a figurative precipice to enter into positive change. Often it takes a disaster for wonderful things to happen. You have to have fire to get beauty from ashes. Folks, I wish that weren't the case, but if you look at nature it won't take long to realize that every day something dies so that something else can live. There's simply no escaping it. I understand pain and mourning. I have chronic illness, Dad is dead and I'm disturbed by the new world we live in. I understand that no one wants to do a rain dance when you're in a drought. I get that. Everyone wants to go to Heaven but no one wants to die. But there is something comforting about knowing how nature works. Her greatest miracle… Life comes from Death. Good from the Bad. The great news about the shadowy valleys we go through is that there's light at the end of the tunnel. We'll rise to new heights because of the "fertilization" we get from the decomposition.

That all sounds good but what should a person do, specifically? The problems seem so large and you feel helpless. The good news is the path to positive change is literally in your hands. There's so much out of your control, but you can focus on what Steven Covey dubs your "realm of influence". You can, as Lucas Nelson sings, *"Turn off the News and Build a Garden."* Or in the case of this book, which is about raising animals, turn off the news and tend a chicken! After work, you can skip the sterile gym and opt to roll up your sleeves and get your hands absolutely filthy. Taste a little of that soil even. Don't be afraid to accidentally step barefooted in Holy Cow Squat. It WON'T KILL YOU. In fact, it will make you stronger. Exposing ourselves to the germs of the farm is extremely beneficial for our microbiome, our health. Actually, I caught Joel himself drinking out of his cows' waterer! With excitement he exclaimed, *"This is why I haven't been sick in 20 years!"* He was so excited it rubbed off and I took off my hat and took a sip. Guess what? I'm still here ;) And now, when we need it the most, there's social and emotional support on the farm as well. Especially with animals. Human interactions have been limited but we can still go to the fields and hang out with the livestock. As Joel says, *"None of them is judging me for not wearing a mask, judging me for how I vote, judging me for how I talk or what illustrations I use."* And I'll assure you, they don't require social distancing. Heck, you can even hug them!

We all have a precipice story. Yours may match up with the woes of the world at large, or it may be more individualized. Maybe a mixture of both. The good news is that we don't have to crumble and conform. Problems might be at their height in your life but hang on dear brother, dear sister. Remember we have to fall in order to get up. To learn our lesson. To get better. Fortunately, you've already experienced the bottom, and holding this book is part of your boost back up. You know what you want, but you're not quite sure how to get there. Well, this is your guide. Once you start implementing what you learn you'll find yourself outside breathing fresh air, getting fulfilling exercise, soaking in vitamin D and sleeping better because you're actually tired from physical work.

We all love hearing the epiphany part of the story. But notice the epiphany usually doesn't come while having a grand ol' time at the water park. The epiphanies come as a solution to some significant problems. We rarely recognize the epiphany at the moment. It's usually later on we can pinpoint what really worked to make the positive change in our lives. This time you can recognize it right away. This is your epiphany moment. How can I say this so confidently? As someone who's followed Joel's advice for over a decade I can tell you, he leads by example and a half a century of success. His guidance has led my family and me to create a thriving homestead that provides most of our own food and continually becomes more productive every year.

I don't care what the News says. I create my own news, and you can create yours. Join me with me and the worms and ignore all that and LIVE. Put your head down and milk the cow, weed that garden, build the chicken coop, hug the kiddos, kiss that spouse, and callus those hands. You'll quickly see the world is a Beautiful place and You have a bright future. Why? Because you're taking control of your life and you're finding freedom. You're growing your own food and you've picked up a surefire classic that will show you the way.

Justin Rhodes
Permaculturalist, film producer, author and teacher

Introduction

By Joel Salatin

Back in 1990 when Teresa and I felt like we'd turned the corner and were going to be successful in this farming venture, we had about 25 cows, no hogs, 500 laying hens, 5,000 broilers, no turkeys, and fewer than 200 customers.

That was a far cry from the 10 cows, 2 hogs, 50 chickens, 2 milk cows and brief forays into half a dozen goats, 20 sheep, and a handful of rabbits we'd had for 20 years. I was 4 when Mom and Dad purchased this worn out farm; both worked off the farm jobs to pay for it. Without time or expertise, we ran it like a glorified homestead.

Things had progressed quite a bit by 1990: a full time salary coming from the farm, happy customers, and a miniscule savings account. Life was good. Realize that was already 30 years after our family took stewardship of this place. But once Teresa and I demonstrated that a young couple could maybe bootstrap a living on a gullied rock pile, journalists and sustainable ag conference hosts wanted to hear the story.

Fortunately, my parents enjoyed living below their means and offered me a piece of paid-for land. It was hard scrabble, but it was paid for. Wouldn't we be a better country if every parent left behind a legacy of value rather than a legacy of debt? Teresa and I dug our heels in and as second generationers built on my Mom and Dad's foundation.

A strange thing happened, though, every time I would present and show my Kodak carousel slides (that's what we used before PowerPoint existed). People would clap but then they'd say "that's nice, but can it scale up? I mean, 25 cows isn't much. And 200 laying hens? Really?" People questioned the commercial viability and credibility of our little venture. We didn't even have a front end loader.

Since 1990, steady progress brought us to where we are today generating 20 salaries, the next generation fully in control, a bandsaw mill, 1,000 head of cattle, 800 hogs, 30,000 broilers, 5,000 laying hens, and 1,500 turkeys. We also lease a dozen properties nearby, serve 5,000 families, 50 restaurants, on-line sales and a farm store. We never aspired to this; it just happened while we worked.

Today when I get done with a presentation, the common refrain is "wow, that's amazing, but can it scale down? I only have 5 acres." I always write to the needs I hear, and I hear this request loud and clear. Because I've spent half my life as more or less a homesteader and half my life as a commercial farmer, I can straddle both ends of scale easily.

For me, cow-day formulas and pig pasture specifications are seamless whether you have 2 or 1,000. I'm happy in both worlds. But most people can't translate things that easily. If you haven't been there, it's hard to grasp the concepts, especially when all the examples seem inapplicable.

Anyone familiar with my work knows I'm an unapologetic cheerleader for full time farming. If you really want to farm, then let's figure out how to generate a salary. Forget the town commute and make it happen.

But I'm more aware now than ever that, just like in our family, it took a generation's wealth to create a germination tray where Teresa and I could sprout. Dad passed away in 1988, right when we knew we were going to make it. He saw where we were headed, in the distance, like a hazy silhouette. Mom contributed her salary to the mortgage right alongside him. Sometimes full expression takes awhile to develop, for all the pieces to be in place: capital, experience, relationship.

I'm also quite aware that many times entrepreneurial enterprises spring out, almost surprisingly, from hobbies. This book celebrates both the hobby and the start. If the homestead scale is your beginning for a bigger dream, even a multi-generational dream, wonderful. If it's more a hobby of security and self-reliance, that's wonderful too.

This book is my attempt to take everything we've learned, from our homestead days to our commercial efficiencies, and apply them to small holdings and home-scale production. Once in awhile I veer off into entrepreneur-talk because that's part of my Joelness, but in the main, I'm talking straight to the folks entering this wonderful world of small scale domestic livestock on a little piece of land to grow great food, love their place, and create family legacy. And older homesteaders looking for additional ideas.

Based on discussions with thousands of homesteaders, I'm convinced that if we could encourage a million successful small-holders, we'll inevitably germinate the next generation of viable full-time food and land stewards. You have to crawl before you walk. If you can't handle two chickens you'll never handle a thousand. Husbandry doesn't get easier with scale; it simply compounds the problems.

A little done well by a lot of homesteaders equals a societal food and farm revival. This book is dedicated to anyone whose heart yearns to participate in the practical caress of our ecological umbilical. Whose soul seeks solace in rural sunsets after a day's sacred work. Who understands the importance of personal responsibility securing and nurturing life's foundations. May you find this book inspiring and helpful wherever you are in this noble journey.

Joel Salatin
Summer, 2021

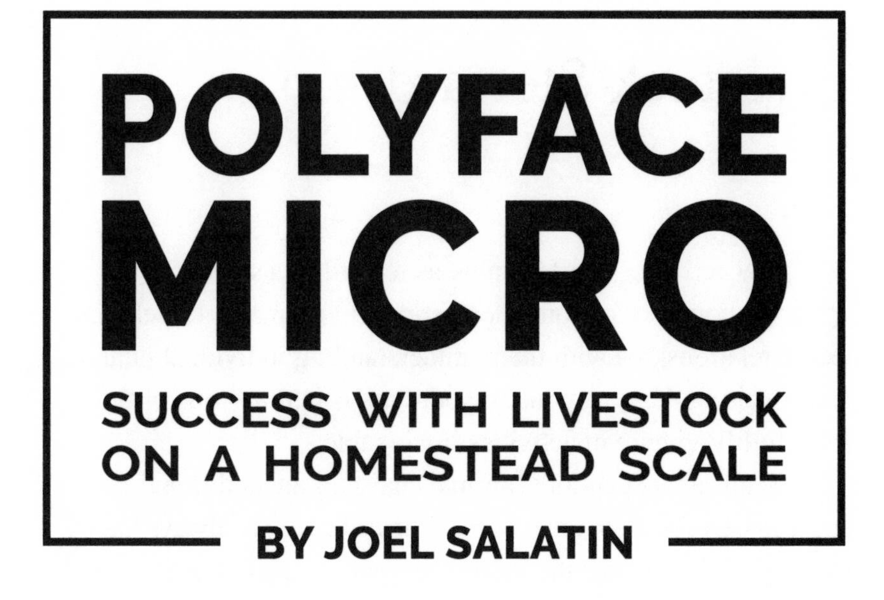

POLYFACE MICRO

SUCCESS WITH LIVESTOCK ON A HOMESTEAD SCALE

BY JOEL SALATIN

Chapter 1

Livestock Stewardship

Few activities in life can be as rewarding as being responsible for some domestic farm animals. Developing relationships with them, understanding individual nuances, and watching them respond to your care is perhaps one of the most fulfilling human activities imaginable.

With rare exceptions, animals have an unconditional appreciation for their caretakers. The milk cow is always happy to see you. No matter the news, the weather, or if your spouse is upset with you, the chickens always run to see what you're putting in their feeder. They're never disinterested or aloof.

Iconic animal behaviorist Temple Grandin says that living in the moment might be one of the biggest distinctions between humans and animals. We humans have a lot on our minds. We think about yesterday, today, and tomorrow simultaneously.

While I'm pouring water in the chicken pan, I'm thinking about whether my daughter enjoyed her birthday party last night and what the commute to the office will be in an hour and if my wife wrote the check to the electrician. Our minds are all clogged up with helter skelter stuff. Not so with animals.

They live completely in the moment. When a cow is in estrous, she doesn't care about relationships or getting in the mood. She doesn't even care if the bull is handsome and she certainly doesn't care if the bull helps her in the kitchen. No, she'll take whatever male with the right equipment gets to her first. Over and done and now time to eat more clover.

While some animals can figure things out, such figuring is relatively rudimentary. Some are certainly more mischievous than others--goats come to mind. And a horse has an uncanny ability to finagle a gate latch. Part of this cognitive function, though, is not really brains. It's just time. While you and I fill out licenses and sit in the auto mechanic's waiting room or cook dinner, our animals have 24/7/365 access to fool around.

And fool around they do. How does a chicken find the tiniest hole in the yard fence? How does the horse jiggle the gate latch until it opens? If you stood there nonchalantly jiggling on that latch for 5 hours, you'd coincidently probably hit the open combination too. This isn't brains; it's just time and persistence.

This time fact has major ramifications. It means that not only do animals love and respond to routine, they also despise and shy away from new things. Remember, as Temple Grandin says, they live in the moment. In her spellbinding and delightful lectures, she talks about wearing a certain hat to do chores. One day we don a different hat.

To us, it's nothing. Somebody gave us a new hat or we decided the old one was too ratty to keep. The point is that we go out to chores with a new hat and suddenly the animals are off balance. They saw the color, the logo, the style every day and suddenly a new color, new logo, and new style enters their world. It turns everything upside down. She tells cowboys to either always wear or never wear sunglasses. If you flip flop, the cows get edgy. You don't want edgy cows.

You'd think the cows would understand that it's the same person in a different outfit, like when you meet people. Nobody would be put off because someone changed outfits in the middle of the day. Animals are far more in tune with what we're wearing, with our look, than we are because their minds aren't cluttered with schedules and other responsibilities.

Animals despise innovation. They want the same thing every day. They don't want vacations, change of scenery, change of schedule. Living in the moment means you don't want surprises. In her corral design work, Temple emphasizes that a tiny fluttering piece of straw on the wall of a chute might as well be an elephant-sized predator to a cow. She's thinking "what is that odd-looking thing silhouetted up there?" Suddenly the cow is tense, edgy, and looking for escape opportunities.

The reason to put attention on this in-the-moment mindset is because if we aren't trying to walk in our animals' shoes, trying to put ourselves in their position, we'll never have happy and healthy critters. The first rule of animal stewardship is to look at their situation and ask yourself "if I were this pig, how would I respond? Would this make me happy? Can I see what my caretaker wants me to do? Is it clear, or am I confused? Does this environment make me happy and healthy?"

This is why in this book I'm going to spend time talking about emotional and physical needs of the animals. I've raised many thousands of animals--perhaps millions--and I still have to remind myself that my first responsibility is honoring the pigness of the pig and the chicken-ness of the chicken. Different animals have different needs and desires. Some, like rabbits, especially dislike loud noises (if you had ears that big, you wouldn't like them either).

Most animals have far better senses than people. They see better (a cow can see about 330 degrees), smell better, hear better, and taste better. Cows eat dessert first--always. They live in the now. Chickens always peck out the tastiest morsels first. Everything is in real time immediate gratification. They don't stockpile for tomorrow. Although bees, ants and pack rats stockpile, this book is about the common farm critters and they have a distinctive philosophy: "you take care of me and I'll give myself for you."

No pig ever stashes some extra tasty morsels over in the corner of the pen to be consumed tomorrow. And no animal holds back to let more timid, smaller, sickly ones have first dibs. All animals are bullies. The pecking order wasn't invented by a bunch of charity minded chickens sitting around holding wings singing kumbaya. No indeed; they're all out for number one, right now, real time, in the moment.

I hope I've established this axiom from Temple Grandin sufficiently to move to more ramifications. The first is that we caretakers must think for our animals. We must think about their safety, their hygiene, their diet. We need to think about where their next meal comes from because all they know is that tomorrow morning the lady in the broad-brimmed straw hat will come and fill my feeder. Who cares how much is in it, how good it is? She will do all that thinking.

We have to see the errant string that the little turkey poult will hang himself on in the brooder. Goodness knows, if we don't see it, while we're at work and that turkey has all day to stumble around the brooder, he'll sure as the world find that errant string and hang himself. I've seen it done more than you can imagine.

What I'm getting at here is that as delightful, fulfilling, affirming, and satisfying as farm critters are, they come with huge responsibilities. Yes, there is a catch. I've got news for you: this isn't all fun and games. It's not all cutesy sweetie pie pet world. When we choose to take the responsibility for our animals' lives, we take on a host of requirements. A milk cow doesn't take Christmas off. The laying hens don't take off the fourth of July. A pig doesn't look forward to fasting at Easter. The only one that consistently takes time off is the turkey, for Thanksgiving. Okay, bad joke.

Our farm animals completely depend on us for their food, shelter, sanitation, and comfort. We don't enter into this partnership on a whim. We don't enter into it because we think animals are cute, or because our wild-eyed Aunt Matilda had interesting chickens. As we begin this journey, as wonderful as it is and should be, we must be serious about how it will impact our lives. When can we travel? Can we go away? Are they too cold? Too hot?

Think about your pet dog or cat when you want to go away, and then multiply that planning and angst by exponential degrees. This isn't a lark; it's a marriage. Mutually beneficial, to be sure, but no less serious. I caution anyone to check your motives. If you're getting an animal because it's cute, wait. Cute wears off the second month. When little pot-bellied piggy gets out the umpteenth time and eats up all your garden potatoes or pushes through the screen door of your house and deposits poop on your carpet . . . well, that cute pig loses some luster.

And don't get animals because it sounds fun, or novel, or cool. None of that has staying power. They'll get sick. They'll run down the road. They'll eat your garden. They'll keep you up at night. Get animals for your homestead because they contribute to the functional ecology and your nutritional security. Trust me on this; nothing else will incentivize you to fulfill your end of the partnership. You'll cheat on your animals (not caring for them adequately) and you might even begin to dislike them. You might wish you'd never gotten them.

The right purpose will protect your end of the caretaking bargain. That leads me to the second major ramification, which is that these are animals. They aren't toys but they also aren't humans. At the risk of irritating animal welfarists, I don't worship my animals. They are there for me, not me for them. This is not selfish and it's not arrogant; it is a fact. While certainly some folks feel a mission to provide sanctuary for farm animals, essentially creating farm animal nursing homes, this is not a practical partnership.

Farm animals offer us tremendous ecological and nutritional benefits that I'll highlight in the next chapter, but their stewardship only makes economic and practical sense if they are productive.

An unproductive animal is parasitic. And yes, therapy animals are productive. But keeping a farm animal beyond its productive life, or for reasons other than production, is simply a drain on the system. Unlike humans, animals do not generally live long after their reproductive capabilities end. A cow that won't breed will soon break down. She'll start limping or lose teeth. Animal virulence goes hand in hand with their end of the bargain.

I make no secret that I'm a Christian libertarian environmentalist capitalist lunatic farmer. My dear friend and author Matthew Sleeth pointed out to me that in all the Bible, no animal is named. Not the donkey that talked to Balak. Not the donkey Jesus rode into Jerusalem on Palm Sunday. Not the horse Absalom rode in his rebellion against his father King David. Not the dove Noah sent from the ark. This perspective is healthy as we examine our stewardship mandate.

While we don't abuse our animals, we don't insufferably coddle them either. They are part of our work team. They must pull their weight. They must pay for themselves. In the end, we as their caretakers must and should take full responsibility for their harvest. And for veterinary intervention. Our animals literally entrust their lives to us; we don't take that lightly but neither do they. It's a fair bargain. We protect them from predators. We fill their feeders. We make sure they have shelter.

In the end, they do not rebel if we mistreat them. Remember, they only live in the moment. They can't conceive of a constitution or a court of law or regulatory oversight. They have no concept of freedom or bondage; they just live with the hand they're dealt. While I believe animals are happier on pasture than in a factory confinement house, from their perspective their situation is unchangeable. Animals can't imagine a different life; they don't read books or watch movies about different living situations. We can imagine it for the animal, and that's why we have to make the hard decisions.

When we cull (eliminate) cows from the herd or turn an old batch of laying hens into stewing hens, we don't do it flippantly. We don't cry, either. It just is. It's part of the stewardship responsibility and allows us to continue enjoying our livestock. If we had to go into emotional counseling every time we culled an unproductive animal or an animal that reached its productive peak (like a grass finished beef or an 8-week broiler chicken), we wouldn't be animal caretakers for long. Some people get too attached. I've shed my tear over certain animals, like a beloved milk cow, for example. But that's not routine.

When people don't grow up around animals with the inevitable birth and death cycle, this weighty responsibility can be overwhelming. If you have animals, they will die, just like people and just like tomato plants or yard trees. Viscerally stepping into this life struggle is not for the faint of heart. You sure don't want to step into it because you think it's cute.

No, enter with eyes wide open. You'll encounter ecstasy and heartache. You'll have animals that respect you, stay in, and endear themselves into your soul. You'll have others that seem like the Devil incarnate, who you can never trust, and who always seem to be cruising toward a wreck. As a livestock steward, you determine the destiny of all your animals. When they come, when they go. And the life they lead under your care. It's a wonderful, wild, wearisome, winsome ride.

Like most things, attitude and perspective set the stage for success. Remember, your pig does not know it's going to be Christmas ham. All it cares about is a clean place to lounge, available food right now, and water. That's about it. Animals are elegantly simple this way. Fortunately they don't attend fashion shows to change costumes or watch TV to see what the other side of the world is doing. This information is timeless. Happy, healthy cows today need the same thing their counterparts did 500 years ago; and 500 years from now, the essence of cow will still be the same.

What I've discovered over a lifetime of critter care I share with complete confidence in its long-term applicability. I love the animals on our farm, both the ones currently productive and the ones in the freezer. God help me to honor that cycle in humility, faithfulness, and creativity.

Chapter 2

The Animal Why

What animals contribute to a homestead is a microcosm of what they contribute on the planet. Consider that no animal-less ecology exists, even at the north and south poles. Indeed, as we've plumbed the depths of the oceans, we now know that even sea creatures exist in the darkest, deepest recesses, thriving under extreme pressure.

If animals inhabit every corner of the earth, a reasonable person would seek to discover why. Are they just superfluous baggage? A transitional encumbrance in planetary history? As anti-animal shrieks grow louder and the cry to ban meat eating crescendos, perhaps it would behoove us all to step back and enumerate the animal contribution to our whole ecosystem.

After all, successful homesteads glean threads of ecological principles to weave a tapestry of resilience and abundance. Let's examine some of the functions animals play in our world and see how they can apply to a smooth domestic operation.

1. Up-cycle the margins. Diana Rodgers, author of *Sacred Cow* and power behind the podcast *Sustainable Dish*, writes as eloquently about this contribution as anyone I know. Many areas of the planet are too remote, too infertile, too inaccessible or too climatically difficult to grow high value crops. In fact, as a percentage of the earth's land surface, only a tiny portion is capable of growing what we would consider valuable plants like corn or squash.

Because animals can navigate difficult terrain--herbivores all have built-in 4-wheel drive--and can find shade or water or walk to food, they can thrive in areas where cultivated crops would be impossible. On most of our homesteads, much of our labor is spent upgrading areas into abundance and that is good. But even after all that labor on these key areas, we still have the edges.

What about that grassy drainage area or the field/forest edge? If you maintain a lawn, that could be considered a non-production area. Why not graze it rather than mow it? A small flock of sheep around fence edges, the outbuildings and other grassy areas will thrive and act like a mobile weed-eater and lawnmower combination, converting those edges into high quality nutrition.

Every place has good spots and bad spots, from a farming and food production standpoint. I'm not suggesting that the promontory that affords the best view isn't a good spot. It's a great spot, but probably not the best spot for your herb garden. The steep hill getting up there might not be conducive to much either, except for honeysuckle and multi-flora rose. But goats would thrive there.

Because animals can walk out and walk back, they can manage on the margins and make something productive out of something that would be too difficult to make productive or maintain with human labor. Realize that all plants respond to pruning. You can over-prune and kill a plant, for sure, but you can also prune just the right amount and stimulate it to healthier growth. When I suggest animals on the margins, I'm not suggesting over-grazing or a scorched earth policy. Quite the contrary; I imagine an Edenic verdant landscape balanced between wild and ordered.

Nature maintains this balance through the predator-prey

relationship, as well as fire, desire for salt, shelter and other things. We'll talk much more about that movement choreography and how to mimic it on our small holdings, but for now, let's appreciate that the bighorn sheep on the crags of the Rocky Mountains provide strategic pruning and the whole ecology benefits from their presence. They are not a net negative; they are a net positive.

The fact that bighorn sheep seem to thrive in such harsh conditions certainly proves the desire of creation to manifest abundance even in the most unlikely places. One of the beautiful things about small acreages is that we can intimately touch every square yard. On our more expansive acreage here at Polyface, I often feel like I need to repent in sackcloth and ashes for the unleveraged areas. I have one acre orchards tucked in various spots . . . but only in my mind. I know where they should go. But with several hundred acres under our management, it's easy to let perfectly usable spots go underutilized due to higher priorities elsewhere.

On small acreages, we can't afford that luxury. Intensification does not mean we eliminate wildness or wildlife. Done right, it often enhances habitat for our non-domesticated plants and animals. What deer isn't attracted to dropped apples from a fruit tree? What rabbit isn't attracted to a carefully piled stack of brush from a clean-up project?

In a small holding, every square yard is precious. Every spot beckons us with opportunities. Near the house we'll no doubt develop gardens, but farther away, we'll use animals on grass, weeds, legumes, vines and even brushy plants to grow milk, meat, and eggs. The herbivore, uniquely equipped to digest cellulose, can thrive on perennials we didn't plant in the margins that we don't see every minute. Gaining productive capacity on these areas, fairly swiftly and without the effort of installing raised garden beds or intricate trellises, puts us on a path of abundance.

If we eventually expand more intensive production into these areas, that's fine. But utilizing what's there, in the condition we find it, to generate food is a benchmark of homestead efficiency. You can do this without planting seeds, without fertilizing, and without worrying about what it will look like in a decade. If you have vegetation out there, even if it's sparse, it'll benefit from strategic pruning, hooves, and manure from animals.

Animals don't care if things are perfect. If a blade of grass or leaf exists, animals will gratefully eat it. Imperfection is just fine. The animals will begin the healing process regardless of how sick the land may look today. The notion that animals hurt the landscape does not come from historically natural and wild ecosystems. It comes from a human history dominated with improperly managed domestic livestock. Over-grazing, as well as over-tillage, has been going on for thousands of years. As judicious caretakers, we don't have to follow that tradition. We can use domestic livestock in its greater animal role to be a partner in healing and not perpetuating the hurt. That is the premise and goal of this book.

2. Up-cycle waste streams. You could argue that this is related to the first point, but it's such a valuable contribution I like to treat it separately. If you visit any pre-1900 farmstead, its distinctive feature is outbuildings. The iconic pre-industrial farmstead had several necessary buildings: a residence, smokehouse, chicken house, pig sty, milking barn, forage and grain barn, buggy garage, shop and equipment storage. To be sure, sometimes these were combined under one roof, or with interesting Ls and tacked-ons, but in general these structures huddled together at the farmstead's hub.

Notice pigs, chickens, and milk were all there, within a few yards of the house. That's because human living requires streams of resources coming in and streams of resources going out. One of the great objectives of a harmonious homestead is to minimize off-site streams. Notice I didn't say eliminate; I said minimize.

Certainly local economies thrive on mutual interdependence.

But one of the foundational mystiques driving the homestead movement is to reduce dependency on "the man." Whether that's the employer, the utility, or the supermarket, the idea is more self-reliance, more of life under your personal control. That entails knowing more about personal living skills, like how to fix things, grow things, and build things. The historic farmstead, then, that existed prior to supermarkets, Amazon, and automobiles, relied on its own resources for the lion's share of inputs.

Farmers didn't go to Lowe's for fence posts. They kept a stand of cedar, locust, or osage orange growing somewhere to provide posts for fences and poles for sheds. They kept seeds from heirloom plants. Indeed, these pre-mechanized farmers devoted a third of their acreage to growing forage for their draft animals. These farms did not get their energy from petroleum or the electric power grid. They created it on their own places through solar energy--chlorophyll and animal power.

They grew big gardens. Big gardens. If they didn't "lay it by" they went hungry in the winter. They preserved a lot of nutrition with fermentation. From sauerkraut to apple brandy, pre-refrigeration and pre-canning, nutrient preservation relied on fermentation. All of these food activities generated waste streams. Pumice from crushing grapes and apples. Rinds from peeling squash and cucumbers. Weeds from the garden. Prunings from the apple trees. Soured milk. Whey left over from cheese making.

Goodness, in those days grain was too expensive to feed as a dietary mainstay to omnivores; it was cheaper to feed skim milk to pigs than to buy grain. This while proximate farmstead living design eventually developed into the legal term curtilage. Today, many states are writing cottage food laws that use the term curtilage to define the extent of inspection and licensing exemptions for small scale in-home value added food products.

The curtilage includes both the residence and the associated nearby buildings and grounds where the family and children practice daily living routines.

A constant flow of animal-edible material moved throughout the curtilage, creating the heartbeat of the family-centric rural experience. If you've ever watched a pig slurp nauseatingly spoiled milk, you'll never forget it. What you and I can't bring within a foot of our noses, these pigs inhale with such gusto it's one of the most pleasurable examples of turning trash into treasure. Who would think that such foul material could produce such fine-tasting hams? But it does.

Meanwhile, the chickens ate every kind of kitchen scrap imaginable. And in turn kept the kitchen supplied with eggs. The ultimate garbage disposal, chickens offered sanitation and up-cycling long before the green political movement. As I'll discuss later, the stationary buildings in yesteryear's homestead we upgraded to mobile structures. With that mobility we've been able to leverage animals as waste up-cyclers far more than our ancestors could have imagined.

Lightweight mobile infrastructure could not be built until pneumatic tires and bandsaw mill technology. When bringing in saw logs required axes and crosscut saws, and the sawmill kerf was 3/8 inches, nobody could profitably manufacture small-dimension lumber. Economics and efficiency demanded large dimension lumber to minimize sawing. Hence, lightweight portable infrastructure was nearly impossible to create.

With the advent of the chainsaw and the bandsaw, we can now make portable infrastructure out of tinker toys from small-dimension logs; this alone revolutionized leveraging animals' waste-consuming abilities. Now we can move chickens behind herbivores to clean up fly maggots. We can move duck shelters through the vineyard to eat up Japanese beetles. We can move pigs through the forest to clean up dropped acorns and mesquite seeds. What a wonderful time to be alive.

3. Manure. What needs to be said about this other than manure is magic? Nothing moves soil to a place of abundance like manure. Composting yields even greater benefits, but anyone who has tried to compost without manure knows how difficult it is. Urban composters constantly scrounge for the nitrogen component. Carbon is everywhere; manure is not. It's precious.

When the anti-animal crowd brags about soybeans being more efficient than cows, they aren't figuring in the whole fertility circle. Soybeans are fine until you run out of fertility. Today's soybeans are still pulling from centuries of manure accumulation. If cows were as efficient as soybeans, we'd have run out of beef centuries ago. Of course cows are inefficient, but that's what enables them to put enough back to build soil.

Agronomists might disagree, but in my view carbon is the most important component of soil fertility. Living things are composed of carbon chains. Carbon drives organic matter, acts as a buffer to pathogens and toxins, and fuels soil biology through decomposition. Carbon and sugar are closely related. If carbon is the soil's fuel, nitrogen is the octane. Nitrogen and protein are closely related. Carbon is light and relatively cheap per pound; nitrogen is heavy and expensive per pound.

The soil needs both, plus many other things, to be fertile. This is a cursory discussion, of course. You can find entire books devoted to carbon and nitrogen. All living substances have a ratio of carbon and nitrogen; the soil needs both. For best composting, you want a Carbon(C):Nitrogen(N) ratio of about 30:1. Because nitrogen is a gas at normal room temperature, it tends to vaporize easily. Keeping nitrogen on the farm and in the soil is a never ending challenge.

Each animal generates a different kind of manure. Because chickens don't pee, for example, they don't separate the more highly nitrogenous liquids from the more carbonaceous solids. That's why we call chicken manure hot. It's only about 7:1 C:N. Cow manure, on the other hand, is about 18:1 in the raw state because more of the nitrogen is in the urine. Rabbit and horse manure are considered the coldest because they're about 25:1 or even 30:1, depending on the amount of grain in their diet. That is gentle enough to be used as a side dressing on nitrogen-loving garden plants like leafy greens or cabbage. It won't burn leaves or roots.

Author and homestead guru John Moody combines chicken manure and carbon scraps to build a foot of garden soil in one year. That can't be done without the manure.

I like to think of an animal as a compost pile on steroids. Digestion is like a compost race track. You can speed decomposition by turning it frequently to aerate it. Or you can just run material through an animal and watch it come out magically in 24 hours. Anyone with composting experience knows that while it sounds simple, it's actually sophisticated. You need the right balance of ambient temperature, moisture, C:N ratio, and microbes to make it go.

But in the digestive journey of an animal, the temperature is always right, the moisture is always right, the C:N ratio is always right, and the microbes are always right. What would take 6-8 months to decompose in a compost pile can go through the process in a day in the gut of an animal. That warp speed decomposition enables us to progress more quickly along the fertility spectrum. Especially for those of us who come to homesteading later in life, we need all the speed enhancers we can get. Animals fill that role.

4. Work. Each animal offers a distinctive talent. Cows can mow cellulose like nothing on earth. That strategic pruning is why all the deepest soils on the planet exist in current or historic prairie ecosystems. If you want to dig deep, pigs are the animal of choice. How about light aeration, kind of fluffing the soil? Then you'd want chickens.

Debugging an orchard? Turkeys are my favorite because they don't scratch very much. They eat three times as much forage as a chicken so they actually mow better. Ducks arguably are even better for this, but if you've ever tried to pluck a duck, you might settle for the turkey as being better overall.

And then we turn to goats. If you have a weed patch or bramble thicket, nothing is happier turning it into productive pasture than a goat. What would take you hours of hacking, chainsawing, and mowing, goats do happily . . . if you can keep them in.

If you're familiar with my family's Polyface farm, you know we use laying hens to sanitize behind the cow herd. The chickens scratch through the cow pats, eating out the fly larvae (maggots), spread the pats to cover more ground, and lightly incorporate the manure into the soil surface. That reduces evaporation of the ammonia (nitrogen) and gets the manure nutrition where the soil microbes can enjoy it.

Our other foundational animal work comes from the pigs, who turn the deep bedding from shed-wintered cows. Rather than making windrow compost piles, we simply let the pigs do the aerating. They don't need spare parts, their oil changed, or a 401(k) retirement program. Substituting appreciating pigs for depreciating equipment revolutionizes the profit potential on a small farm. Everything you can do to rid yourself of things that rot, rust, and depreciate will put money in the bank.

The beauty of letting animals do the work is that rather than just being seen as egg producers or pork chops running around, they become co-laborers with us in this great land healing ministry. They are partners, part of our team, and that adds an exponential emotional and spiritual value component to their existence. I'd like to know how folks who think domestic livestock should be outlawed expect us to perform all these necessary functions without animals. I'm confident they've never contemplated the functionality of an Eggmobile or the sheer ecstasy of pigaerators.

One of the slickest work animal set-ups I ever saw was at

long-time draft-power produce growers Eric and Anne Nordell's farm in Pennsylvania. They shifted their work horse stalls every day or two and had a network over the stables that looked like a glorified HVAC (Heating, Ventilation, and Air Conditioning) duct system. Except the ductwork was chicken wire. They had a chicken coop at one end of the horse stables. The chicken wire tube ran above the stables. At every stable, a door could be opened.

Each morning the chickens would awaken and head up into the tube to find the open door into a particular stable. That way the Nordells, via the chickens, could stir the bedding, eliminate flies, and harvest bits of grain the horses passed. The rotational stable sanitation system worked perfectly and yielded beautiful eggs from a process that otherwise would have cost the farmers substantial effort.

High tech treadmills are now becoming a thing. A goat on a treadmill to operate a washing machine or water pump, for example, is now possible with high tech machining. Don't underestimate the work animals enjoy. All you have to do is honor the distinctive proclivities and instinctual desires of the animal and you'd be amazed what these partners can do.

5. Food storage on the hoof. In most of the world, wealth is still measured in livestock. Why? Because where socio-political unrest rules and where refrigeration doesn't exist, nothing beats animals to provide real time food security. When people flee, they don't take watermelons. They take a basket of chickens, a cow, or a goat. Many an American pioneer saved the children by getting them a cup of fresh milk a day from the cow ambling along behind the wagon.

For the Native American, bison were a walking supermarket. Bones for tools, sinew for rope, hide for shelter, flesh for nutrition. Everything they needed came from the bison. When reflecting on the beauty and resourcefulness of such an existence, don't forget to appreciate what the animal brought to the culture. Historians tell us that when hunting was easy, Native Americans would eat as much as 10 pounds of bison in a day. They couldn't always be sure about tomorrow's meal, so they learned to feast when they could.

Animals offer nutrition in real time. It doesn't require refrigeration, canning, or carrying around. Animals can move with you; they are the ultimate real time food security. A mouse or rat won't come and eat your cow while she sleeps at night. But vermin will steal your beans and rice. The cow doesn't get moldy when she's two years old. Dry goods can and do get moldy over time. All foods lose some of their nutritional punch once harvested.

This is one reason why the chicken is such a preferred homestead animal: it can be eaten in one sitting. If you're going to eat a cow in a day or two, you'd better invite all the neighbors. But a chicken comes in a family size, a Goldilocks package, that's juuuuust right. Historically, one way to extend an animal's edibility time frame was to keep it hot, like in a stew. Hence, Brunswick stew became popular in Great Britain as a fortnight (two week) option for the hunt camp.

Hunters would bring their small animals like pheasants, rabbits, and squirrels to the simmering stew pot. The heat kept anything from spoiling and provided a bowl of nutritious victuals whenever a group of hunters returned from their escapades. It was the ultimate slow food fast food.

Whether you view the animals as an ongoing stream of on-the-hoof food or as an apocalyptic security option, their unique ability to hold food against the elements is a real benefit. Perhaps financial guru and best-selling author Dave Ramsey, founder of Financial Peace University (FPU), could add a food component to his six-month emergency fund. When calamity strikes, like Covid-19 in the spring of 2020, and the supermarket shelves go bare (remember, meat and poultry were the first things gone), an emergency fund of cash in a bank won't put food on the table. A larder and some backyard chickens will. That's the kind of stuff homesteaders think about.

6. Personal nutrition. I won't belabor this too much because I know folks willing to read this book don't need to be convinced about the dubious nutritional and pathogenic qualities of corporate food nutrition. In this point, I simply affirm what folks in our homesteading tribe have long believed: factory farmed meat and poultry are neither nutritious nor safe.

The nameless, faceless industrial animal system probably never did care about personal nutrition. The mechanistic mindset sees life as fundamentally inanimate. Biology is physics and chemistry; no more and no less. That gives license to manipulate and adulterate to the extent of human cleverness. Freedom to do what we want rather than what we ought has led to a whole modern food linguistic nightmare, from superbugs to food allergies to highly pathogenic *E. Coli, salmonella* and *camphylobacter.*

Nature, through these new words, screams "Enough!" but we swashbuckling techno-nerds drunk on hubris continue inventing more sophisticated concoctions of Devil's brew and Devil's protocol. Those of us who want to look at our food in the eye, who want to nurture it BEFORE the plate, who want to create personal authenticity in our sustenance, yes, we realize the value of knowing, seeing, holding. Visceral accountability starts long before the frying pan. It starts out in the brooder, out in the field where life springs anew and our human responsibility caresses all the way to the kitchen.

More and more of us feel overwhelmed by the things in our lives we can't control. We can't fix our cars; we can't fix our refrigerators; we can't fix our computers. We don't know where our water comes from, our power, or even half of the ramifications of our investment portfolio. Our inner souls cry out for a handle on something, on anything. What can we know? What can we steer? What can we serve? Answering that cry is the bleat of a lamb, the gentle nudging of a cow coaxing her newborn to stand and nurse for the first time. In the miracle of birth, we find satiation of spirit and sustenance.

When we sit down to partake of our repast, steaming meat loaf or honey baked chicken (two of my favorites), we can know the provenance, the provider, the production techniques. Indeed, we can personalize our nutrition.

7. Emotional conviviality. Finally, we come to the esoteric part of all this. I've saved it until last not because it's least important, but because it's the capstone of everything else.

Anyone who has a people-centric farmstead knows that animals are magic. If you're hosting school tours or city cousins, children will abide the garden tour and enjoy the plants, but they'll dance at the animals. Anyone familiar with my work knows that perhaps my most favorite thing in the whole farm is to go out on a summer evening, at sunset, and lie down in the pasture with the cows.

By sunset the herd, which was probably moved onto the new paddock around 4 p.m. is now full and literally lethargic after feasting. The satiated, curious beasts ease over to my prone physique and gather around like people at a campfire. Inquisitive, but not wanting to get hurt, they study this strange form like a doctor studying a patient. Finally, and usually this is within a few minutes, one of them steps close enough to sniff over me. I feel her hot breath spurting from her massive nostrils. The gentleness exhibited by these massive animals is probably the most exhilarating part of this experience.

A cow could easily place her hoof on my chest or face and crush me to death with a single step. Imagine the scene. I'm surrounded by 20-25 cows, each weighing 1,000-plus pounds; they stand; I lie on the ground. But in decades of doing this with thousands of animals, I've never had one come close to stepping on me. The few who don't like me don't come close. The friendly ones are just curious, sniffing, then rubbing their whiskery muzzle over my body. My favorite is when one reaches out that foot-long sand-paper tongue and licks my cheek or my britches. If it's a hot day and I've been sweating, they no doubt get some salt for their efforts.

I invite anyone to come and do this. I never tire of this experience. It puts me in the proper posture before these wonderful animals. I'm completely vulnerable, in their space. And all they want to do is touch, smell, look. That's enough. They stay as long as I stay. Some wander away from the circle and others take their place, eager to see what the fuss is about. Again, I'll simply say to the folks who think domestic livestock should be outlawed, who think that those of us who keep them are nothing but exploiter Conquistadors, worse than Satan; I implore you, come and lie with me, in the field, and let the cows caress you. It's the best way I know to portray the relationship that animals bring to a homestead.

This is different than a pet dog or cat. I've been accused by rabid vegans that if I would eat a cow I'd probably eat my cat. Well, I would if I were starving. But I'm not starving and I don't plan to eat my cat. This domestic livestock relationship is different than the one you have with your pet. It's not a pet. It's not better; it's not worse; it's different. And that's okay. But it's still special and touches us emotionally beyond the reach of a cabbage or tomato. And I love cabbage and tomato; I don't mean to slight my garden buddies. It's just different . . . and special.

I'm not a psychologist, but I submit that few things offer emotional therapy on the level of farm animals. Their unconditional loyalty and affection is almost unparalleled in the human sphere. Our animals need us and what's more, they want us. And they appreciate us. Although these days I don't move the pastured broilers as much as I did years ago, I still enjoy it, a lot. Why? Who gets to wake up every morning and make this many beings happy? My emotional cup runneth over. And so can yours, with animals on your homestead.

Chapter 3

Biggest Pitfalls

Now that you're all excited about the animal asset and opportunity, let's dig into things that can reduce all these benefits. This is not a downer chapter; it is a be forewarned chapter. Be assured that I've been guilty of every one of these pitfalls. I've been at this long enough to make every mistake you can imagine.

If you will commit to learning and appreciating these pitfalls, your experience with livestock will be wonderful. If not, it'll be a nightmare. The reason most homesteads have a six-year cycle is because these pitfalls turn enthusiasm into depression and then into burnout. Two years is the honeymoon, then two years is the new realization that things aren't as easy as they seem in the books, and then the last two years are the "For Sale" period. While I want to inspire you in an animal direction, I also want to shoot straight about the requirements for success. Dismissing any of these pitfalls will cost you in time, money, and joy. Here we go.

1. Control. By far and away the biggest pitfall is inadequate control. Being able to put the animals where you want them, keeping them there, and then being able to move them to another spot efficiently is the bane of homesteaders. Again, too many come to this farming thing from a pet background.

Assuming that since your dog comes when you call does not mean Bossy the cow will come when you call.

If Bossy came from another farm and is a new acquisition and has never seen electric fence in her life, she likely will jump off the trailer in your field, run through the fence like a hippopotamus, and head out to the highway. The ensuing mobilization from State Police, Sheriff's deputies, local farmers and fire company volunteers to get Bossy corralled will be a migraine memory. Been there, done that. One fall we actually lost three stocker calves. They went through several electric fences, headed up the mountain and we never saw them again. I have no idea who eventually benefitted from their presence, but it sure wasn't us. A nice cool loss of $2,000. And we thought we knew what we were doing.

My intent here is to scare you to death. I want you just short of terrified. Almost every newbie homesteader I know has a nightmare control story. Cows are the most critical because they are the biggest and can cover the most ground fastest if they get away. But pigs aren't far behind. And they can do a lot of damage in the neighbor's flower garden. Been there, done that too.

I will discuss fencing later, but I need you fearful enough to pay attention when we get into fencing details. Too many folks glaze over fencing as if it's a secondary issue. It's not. I beg you; if you want animals, securing them where you want them, when you want them, and being able to move them is foundational to success.

Control includes preventing your dog raising havoc with the animals. A dog that chases sheep or chickens is a common homestead hazard. I'm a huge fan of Cabella's shock collars for dog training. They extend your reach and the dog can be disciplined without knowing you're involved. Over the years we've had a lot of cats for vermin control, but sometimes they get a bit too aggressive on chicks. A whiffle bat works very well in training a cat. The light bat is strong enough to get attention with a whack to the rear end, but not enough to inflict bodily harm. Corporal punishment works extremely well if done with the right attitude, at the right time, and at the right strength. Any pet that agitates your domestic animals is a liability. Watching a dog chase chickens is not funny.

Securing your vulnerable small animals from predation is also part of control. Remember, predators have all the time in the world to inventively invade your poultry quarters. Portable coops can be less fortified because they're moving, which keeps the predators a bit off guard. But if you have a stationary coop, it needs to be a Maginot Line. Every night predators come for a visit, and the hungrier they are the more aggressive they become. Stationary coops offer compounding digging and yanking access to predators, whereas mobile ones require a new dig spot every night. Predators tend to be opportunists; even slight disruptions often dissuade their efforts.

Perhaps the most overlooked control element relates to movement. How easy is it to take a flock or herd from one spot to another? Movement does not always require physical fences. My grandson has a small flock of sheep that are bonded to their guardian dog. With the dog on a leash, Andrew can lead that flock from one end of the farm to the other; they just follow the dog. To be sure, that level of control took proper training of the dog, proper bonding, and diligence to keep the dog with the sheep, but it allows this flock to be controlled without changing all the fencing on the farm.

Finally, what do you do with a rogue? Sometimes a particular animal won't fit. It's always testing, pushing, trouble making. You can try behavior modification like solitary confinement and a short length of hot wire across a corner of the pen. Andrew had a ewe that figured out how to jump the electrified netting. We tied a piece of firewood to her neck and let her drag it around. She could eat fine and get around, but couldn't jump the fence because the stick of firewood weighted her down. After a month, we removed it and she never jumped again.

Sometimes, however, an animal will not abide re-training. It won't respond to anything you do. That one needs to be culled. Putting wheels under that animal and sending it down the road will be a happy day. Don't put up with incorrigibles; they'll suck you dry emotionally and financially. Greg Judy, guru of multi-species grazing, developed his easily controlled sheep flock by religiously culling the ones who escaped.

He began culling years ago for everything from hardiness to fence jumping. If he found a sheep outside his electric fence, he got rid of her. This ruthless culling regimen resulted in perhaps the most disciplined flock in America; people come from all over the country to get his well-disciplined sheep. These traits are highly heritable but can only be cultivated with strict standards and aggressive selection. If you name your sheep, you simply won't practice this level of excellence. When Justin Rhodes, high priest of homesteading, tired of trying to keep his sheep in, he actually bit the bullet and bought sheep from Greg. He told me he thinks these Greg Judy sheep would stay in with only a strand of baling twine.

The thing to remember about control is this: the animal is only doing what it naturally wants to do. Don't blame the animal. If I have an incorrigible animal that no amount of behavior modification seems to affect, the problem is I bought the wrong animal. It's never the animal. We're the ones making the decisions; the animals are simply responding to our decisions. If we decide to laugh at the dog chasing the chickens, we can't blame the dog next week when he kills one. The dog is just being a dog.

As the caretakers, we must be the adults. We must do whatever it takes to have our animals under control. Sometimes that means liquidating and starting over. Sometimes it means getting rid of a pet. Sometimes it means being more clever than the animal. All animals can either be controlled or sold. The number one pitfall on every homestead I've seen is lack of control.

2. Water. The three greatest technological advances in farm animal management, in my view, are electric fencing, chainsaws and sawmills, and water pipe. Plastic piping is cheap and available. You cannot afford to haul water. For emphasis, let me say that once more: you cannot afford to haul water.

Water is heavy and it sloshes about. Waiting for a vessel to fill and then hauling the heavy trailer is time consuming and rut inducing. You can tote a lot of water in buckets, quickly, as long as the distance is short. A nearby spigot changes everything. Even inch and a quarter 100 PSI piping costs way less than a dollar a foot. That means you could install a mile of it for less than $5,000. Think about what folks spend on homestead infrastructure and I don't think you can show me any other investment that can pay off as rapidly and multi-functionally as water pipe.

In a later chapter I'll drill down on water a lot more, but for now I want to establish the point that you cannot afford, either in time, money, or wear and tear on roads and equipment, to be toting water. One hundred gallons of water weighs 800 pounds. Five cows on a hot summer day will drink 50 gallons (10 gallons apiece). Even if you have a hydrant that can deliver 8 gallons a minute, it'll take you 12 minutes to fill that 100 gallon tank. You have to haul the tank in from wherever it is and then take it back to the field. That's another 15 minutes. That's 27 minutes for the 100 gallons.

What's your time worth? If you assume your time is worth $20 an hour, that's about $10 per trip or 10 cents a gallon for water. If you add in wear and tear on equipment, and fuel, it goes higher. If that happens every other day, you'll have 180 trips a year at $10 each is $1,800 plus depreciation and fuel, which means you've spent $2,000 in a year just toting water to 5 cows. That just increased per-cow keeping costs to $400 apiece. That means when the cow calves, you can't make a dime on the first $400 in calf value.

Compare that to a one-time capitalization cost of 2,000 feet of water line at an estimated $2,000. That one-time investment will last for 100 years. (I don't know how long black plastic pipe will last when buried, but it's a long, long time). That's an annualized cost of $20 per year plus a tidbit of electricity. Again, we'll address water sourcing and logistics later, but please believe me in this pitfall chapter: you cannot afford to haul water.

Go ahead, I dare you to pull out your wish list of homestead infrastructure. You might have tractor, trailer, chipper, shed. Yes, they're on mine too. But I think I can guarantee that a dependable water system should precede everything on that list. You can't have life without water. And you can't really thrive living on your homestead until you can walk every corner and enjoy pressurized, fresh water.

You need on-demand pressurized water distributed throughout your homestead; delaying that investment simply eats up your time and money. Most homesteaders say they don't have enough time or money. Other than chasing out-of-control animals, toting water is the next biggest pitfall.

3. Poor genetics. Next to my dad, my greatest mentor was Allan Nation, founder of *Stockman Grass Farmer* magazine. I wish I could remember all his pithy wise sayings; he had one for almost every situation. One I remember was this: "If you go to a sale and in 5 minutes can't figure out who the sucker is, it's YOU." He'd always punctuate the YOU with a mischievous grin.

I'm going to apply that to this pitfall. You and I would like to think that a newbie farmsteader coming into a community would find the old codger agrarian pillars eager to help an aspirant get started. Some do, and that's great. But others don't. They'll sidle up to you, act real friendly, act like they know every critter and good deal in the area. But they really are just predators on your ignorance and they'll take you for a ride.

They're looking for suckers. Don't let it be you.

Lots of homesteaders get talked into buying animals that are somebody else's problem. Every farming community has bottom feeders who operate in the nefarious trading circles as salvagers. They look for the odd deal, the misfit, the one-banger bull calf, the part-Holstein black calf. They'll sell a 3-quarter milk cow and a worm-infested sheep when they can.

Just like you'd vet a spouse, put due diligence on vetting your source of animals. Never be in a hurry to buy, and never buy from someone in a hurry to sell. This is not a TV transaction with the "but wait, if you buy right now we'll throw in a second one for free" deal. Ask around and invest some time in references and scuttlebutt. You want to do business with someone whose operation looks like the one you aspire yours to look like.

The key is to find compatibility. Sometimes poor genetics can't be seen in the animal. A little grain can cover up a lot of problems in an herbivore. If you're wanting to produce herbivores (cows, sheep, goats) without grain, then for goodness sake don't buy animals from someone feeding grain.

Disposition is also a big factor. You don't want high strung animals. When you go visit the potential seller, watch how he interacts with his animals. When he walks through them, are they calm? Is he gentle? Disposition is an extremely heritable characteristic.

Perhaps this pitfall is most commonly seen in laying hens. Inevitably somebody sees chickens on Craig's List for $1 each. What a deal! But they're two years old and infested with mites. Laying hens only produce well for two years. You get your deal of the century home only to find that they've brought all sorts of problems to your homestead, not the least of which is that they won't lay very well since they're already old. Next thing you know you're heating up the scald water for a day of butchering these tough old ornery biddies.

Cast-offs circulate through every farming community and if

you're not in the fraternity, you'll be the sucker who bites. Vetting is a skill just like any other; the more you do, the better you'll be. The more farms you visit, the more observant you'll be. Mastery takes time and usually comes with a few mistakes.

In full transparency, my own reputation in our farming community is suspect. Some commercial farmers in our area call me a Typhoid Mary and a bioterrorist. Just because someone says Farmer Joe is a bad hombre doesn't mean it's necessarily so. That's where being eclectic and talking with people whose outfits look like the one you'd like yours to look like is important. Just like you want to select partnerships that fit your community, character, and culture, you want animals that fit your values as well. Many commercial farmers in my county consider me a pariah because I don't vaccinate, medicate, poison my fields and apply chemical fertilizer. Sometimes the badmouthed person is really the innovator and best steward in the community.

The most common reason why people succumb to poor genetics is speed. "I'm on my homestead and I need animals right now" is the fastest way to a train wreck. Take your time; be eclectic and diligent in your sleuthing; find compatibility. Interestingly, the poor genetic pitfall leads us straight to the next one.

4. Exotic genetics. This is the flip side of the previous coin. I'd like to know how many billions of dollars homesteaders have lost chasing exotic genetics. From ostriches to heritage breeds, this pitfall swallows up folks by the hundreds.

My own foray into this was years ago with bulls. Since I'm a slow learner, it took me two times to understand the lesson. Back in the 1980s, I decided I could upgrade my cow herd rapidly with AI (artificial insemination, not artificial intelligence). Just so you know, never confuse barnyard lingo with geek lingo. It'll get you into trouble.

Just so you know, in the hotshot AI bull stud industry, technicians electronically stimulate bulls to ejaculate semen.

Generally, one ejaculation creates 100 service straws. A straw is simply a pipette carrying the semen and it's squeezed through the cervix of the cow for impregnation. Stored in liquid nitrogen, these straws have a multi-year "use by" date. It's quite remarkable and a testament to technological ingenuity.

I learned how to do AI and got a semen tank, looked at the bull registry, picked my two pretty boys and got some straws. I bred several cows and four of the subsequent calves were heifers. I used natural service on them when they finally got big enough and old enough to breed. Two of them would never breed. Only one of the two that bred eventually calved without assistance. She had one more calf and then would not breed again. That's a pretty dismal average. Overall, that was one of my more expensive lessons.

Not to be deterred, I purchased a bull from a state experiment station that had been "on test." That means seedstock producers put their animals in a pool measured for performance. When we brought the bull home, he was pretty alright. But the next spring, I pulled about 75 percent of the calves. They were too big for the cows to birth without assistance. By that time, I was a bit older and wiser and realized that hot shot genetics were not compatible with our hands-off, no-grain low-cost system.

Today, we buy bulls from outfits that don't use any grain, that do controlled grazing, and control with electric fence. We'd love to add no chemical fertilizers to that list, and I think within a few years we'll be able to. Just like every farm community has the salvage operators, it also has the hobbyists that have lost a pile of money and look to newbies to help bail them out. These folks entice you with clever messaging about rare breeds or organic certification to lure you into patronizing their overly expensive ideas. Don't fall for it. You're under no obligation to make them feel better about their luxury decision.

Messaging comes in all forms. Let's do a bit of math. Suppose you buy four bred heifers from a rare breed enthusiast for $2,000 apiece. Mind you, local regular heifers of the same age are available for $800. But this fellow convinces you that this rare breed is special because its heritage goes back 800 years to a hamlet in Scotland where Vikings from Norway offloaded the original stock when a botched raid required a quick exit. Ahhh, what wonderful nostalgia, connection to the old country, a piece of history right here on our homestead. Yes, bring me four please.

You now have $8,000 tied up in four animals that will calve in six months, let's say. Calving time comes and you watch them for the tell-tale signs of imminent calving: a full udder and swollen, loose vaginal area. Three of them calve fine but one loses her calf in a vet-assisted difficult delivery. Now you have a $200 vet bill and three calves. If you assume a 10-year life on those heifers, their annual depreciation is $200 per year. But it's far more than that because on average, at least 25 percent of heifers will be dysfunctional. You just found out which one was dysfunctional. So really the $8,000 is spread across only three, which bumps the annual per-cow depreciation to $266.

That means every calf costs $266 just in maternal depreciation. I won't belabor the journey, but be assured that another of the four will not make it past five years. Only two will make the full 10 years. Now, what do we do with that $2,000 bred heifer we fed for six months and spent $200 on a vet bill? She has produced nothing and our head screams: "get rid of her!" But then our heart kicks in, and we think about all that $2,200 plus care and feeding and well, maybe if we gave her another chance and tried again . . . ."

Trust me, this is the bane of all animal caretakers, not just homesteaders. But because small operations have fewer animals, the emotional attachment is far more severe than it is for someone who has 500 cows. Bigger outfits know that every year they'll cull 10 percent of their herd. Let's assume your head wins out and you do what you know you need to do: put wheels under her. So you take her to the sale barn and alas, she brings $800. Wow, talk about a way to lose money fast.

Now contrast that with four bred heifers you bought from a reputable neighbor, regular, run-of-the-mill genetics, for $800. Same scenario but this time, when you take that culled animal to the sale barn, she brings you $800. You haven't made anything, but you haven't lost your shirt. Realize that my conservative scenario here is often skewed far more than I've presented. I know folks who bought bred heifers for $3,000 apiece. Imagine what that would do to the above example. If you add hauling water at $400 to the $266 cow depreciation, you're suddenly at $666 burden on that calf before you can begin to break even. And we haven't fed the cow or the calf yet. It's easy to lose money in farming.

Here's my advice: always buy regular genetics from a reputable outfit in your community or as near in location as practical and upgrade with the bull of your choice. Put your money in the bull, not the cows. You can usually sell a bull for what you paid; the same is not true for cows and heifers. As long as he's sound, a bull holds value. The females are far more depreciation risky. Getting more common genetics and then covering with the sire that heads you in the direction you want to go might be a slower trajectory than you'd like, but it's much safer economically.

Another problem with exotic genetics is processing costs. Abattoirs tend to charge the same amount to slaughter a big steer as they do for smaller steers. Ditto hogs. If the kill fee is $75, it makes a big difference whether that is spread over a 500 pound carcass or a 300 pound carcass.

In addition, many exotics grow far slower. For example, the grass-based hogs like Mangalista or Kune-Kune grow far slower than more conventional breeds. These create a double whammy: they're small carcasses and grow slow. If you value your time at all, think about what an additional year means to the price you have to charge to compensate. An additional 10 hours per hog in production, plus the prejudicial kill fee, can easily require $200-$300 per hog. If you have 150 pounds of meat, you might need an extra $2 per pound just for the privilege of having exotics. Cute and unusual don't pay the bills and don't necessarily increase functionality. Common often is common because it works.

5. Pet mentality. Unproductive animals are a massive pitfall on any homestead. Every animal must carry its weight. These are not pets; they are production animals. You don't get goats because you need something with big ears to rub. Certainly you will want to rub their ears, but that's not why you have them.

The stockman's rule of thumb for culling is the three O's: open, ornery, and old. Every breeding female needs to birth on schedule every time. That said, one of the special opportunities in small scale animal production is that often you can overrun the commercial averages due to lower competition. An animal fighting for the water trough or feed box has a much easier time of it when only a couple of other heads or beaks are there. If it's 1,000, the stiffer competition inherently decreases productive longevity.

Certainly a laying hen will lay past two or three years, but she is more prone to sickness and won't lay enough to pay the feed bill. If you're going to mess with the chickens a couple of times a day, you may as well mess with productive ones instead of unproductive ones. As a rule of thumb, whatever the commercial average longevity is, you can figure at least 30 percent more in a well-cared for small group.

Dawdling over unproductive animals is frustrating. If you have 10 sheep and one of them is sick and needs medical care, you can easily spend more time with that one animal than all the others combined. Before you know it, the whole operation suffers because of one. Obviously the more valuable an animal, the more you want to try to save it, but in my experience, most heroic measures never pay out. If one in a group gets sick, I view it as a blessing that the weak individual is identifying itself.

If I blanket the whole group with vaccines, wormers, medications and other interventions, I may never know who the weak individuals are. The only way to determine the weak ones is to kick out the crutches and see who stands.

Tom Lasater, who developed the Beefmaster breed, required every cow to wean a calf every year on schedule. Somebody asked him one time if he would cull a cow whose calf was struck by lightning. Without hesitation, he is said to have responded about like this: "Absolutely; she shouldn't have had her calf where lightning would strike. Not having to raise that calf gave her paid vacation which was unfair to the others who put in a full year's work."

The only way I know to move forward with adaptability is to cull ruthlessly. The more second chances and interventions you offer, the more second rate animals you'll have in your herd or flock. Over time, you'll develop a group of highly dependent animals rather than a group of independents. The one thing you don't want is more dependence.

The most dramatic healing I ever experienced was a cow that became paralyzed during birth. I took her some hay and kept a bucket of water next to her head. She was perky so I resolved to do minimal care and see what happened. After a week without progress I went to my old-timer neighbor and asked him how long I should give the extra care. He said to give her 30 days.

I kept giving her hay and water for another week. Then another. By this time, I was marking the days off on the calendar. Another week. The thirtieth day was a Sunday. We went to church and when we came home shortly after noon, as we drove in the lane, guess what we saw out in front of the barn? That cow standing up! Thirty days to the day. I tell this story so you'll know I have a lot of heart too. It's okay to let your heart rule your head once in awhile.

One old cowman told me his procedure for a paralyzed cow was to pick it up in a front end loader and take it to a pond. He dumped it in and if it were ever going to get better, the shock of the water and will to survive would kick in and it would thrash its way out of the water. If it wasn't going to survive, it would drown. Sounds kind of cruel but perhaps it would beat a month of carrying hay and water to nurse a cow that wasn't capable of recovering.

A pet mentality will wind your heart up in circles and drain your bank account. Resolve today to keep only productive animals. Erring toward extra culling, in the long run, never hurts as much as lenient culling. I've never cared about colors, sizes, or pedigrees; the only question is "are you productive?"

6. Improper diet. All animals have a preferred diet and they all eat dessert first. Remember, they're in the moment, so they don't know how to leave some tasty morsels for later. They always eat ice cream first.

Not only do dietary needs vary from species to species; nutrition requirements change with maturity. Generally, babies need high protein. As animals mature, their protein needs give way to starch. They always need some of both; I'm speaking in general terms.

For mammals, milk offers that high protein diet. Nursing gradually gives way to coarser feedstocks and more carbohydrates. Finishing animals need high energy. Omnivores go through the same stages. In the industrial poultry industry, the carbohydrate:protein ratio in the feed ration changes weekly. Protein is expensive; keeping it to a minimum saves feed costs. On a homestead, you don't need to be nearly as precise. In fact, in our pastured broiler enterprise we don't fluctuate the ration for the entire life of the bird. Since these fast-growing chickens eat supplemental pasture, they're getting more than what's in the feeder.

In a confinement factory house, the ration is far more fragile because the only thing the chicken can eat is in the feeder. In a pastured poultry operation, that is not the case. The diet variety builds some forgiveness into the ration. Ditto for turkeys, laying hens, and hogs--all the omnivores.

Perhaps the singular biggest deficiency in both small and large scale livestock is adequate mineral. Remember that in the wild, animals travel long distances to find salt licks. Historically, fires offered mineral-rich ashes and charcoal, attracting animals for miles around. Native Americans actually used fire as a wildlife attractant. But our domestic animals don't have the luxury of traveling overland to find mineral-rich ashes and salt licks. We must provide that for them.

I'm a huge fan of kelp, and my favorite brand is Thorvin from Iceland. It's geothermally dried which preserves the highest number of cytokinins (hormones) as well as the minerals. Although we raise many thousands of animals every year, our vet bill is zero. Of course we lose an animal once in awhile, but so do folks with big vet bills. We'd rather spend money on good minerals and not fool with vet bills than be miserly with minerals and deal with deficient animals.

Just because you have a small acreage and a little flock or herd doesn't mean your animals will be thrifty. If your herbivores are cribbing (chewing on wood) that indicates severe mineral deficiency. Invest in good minerals--not the artificial stuff at the feed store. Patronize the advertisers in *Stockman Grass Farmer* or *ACRES USA*---join the tribe. You want compatible businesses that think like you think. If you dislike artificials, then don't buy artificials.

All feedstocks are not the same. A common misconception is that all hay is the same. It's not. Hay can be over mature, moldy, or a variety the animals may not like. If you're unsure about the hay quality you're buying, get a sample bale from several sellers and put a flake of each in front of your animals and watch them. A lot of successful animal husbandry has to do with taking the time to watch the animals. Since they don't watch TV and see advertising or celebrity endorsements, animals give unbiased audits of feedstocks.

Each animal knows all it needs to know to be a successful animal. I've watched cows literally burrow through inferior hay to get to better quality even if it's buried in the bottom of the hay bunk. Trust the animals' choice; they will never lie to you.

High performance animals need high octane rations. Cornish Cross broilers can't get along on 14 percent protein. Turkey poults need about 28 percent protein, like game birds. When I have questions about rations, I call Fertrell in Bainbridge, Pennsylvania. I also consult my Morrison's Feed and Feeding book; it's a compendium of feedstock information and will enable you to formulate your own ration if you want to. Just be aware that feedstocks vary and animal requirements vary from species to species and depending on maturity. One size does not fit all.

7. Poor sanitation. Perhaps it's too harsh an indictment, but I'm going to say it anyway: some of the worst animal sanitation I've seen is not on factory farms, but on small-scale homesteads. Hygiene is not about scale, although I admit all things being equal it is easier to keep things clean on a smaller scale.

Hygiene is about proper sanitation protocols. I'll devote a whole chapter to this topic later on, but when we're discussing homestead pitfalls, this needs a mention. If your animal quarters stink, you have a sanitation problem.

Your children should be happy to go eat a sandwich with your animals, regardless of season, regardless of whether they are protected indoors during a blizzard or free to roam a pasture in the summer. At any time and regardless of circumstances, your animals should be aromatically and aesthetically sensually romantic.

Andy Lee, author of *Chicken Tractor*, used to say that if he took a visitor on a tour of his place and had to explain away something that wasn't right more than one time, he wasn't taking care of business. Anyone who's had animals knows that occasionally something isn't quite right. But if it's routine, you have a maintenance problem. You should be able to take surprise visitors around your place with confidence, knowing things are presentable and attractive.

Here at Polyface, we're famous for our 24/7/365 open door policy for anyone from anywhere to come anytime and see anything unannounced. We've learned such a policy is worth it just to keep us on our toes. Have we ever been embarrassed with such a policy? Absolutely. But having to apologize about something makes us aware of the issue and forces us to double down on a solution. Too often folks just assume barnyards are smelly.

"Oh, it's that fresh country air," we say cavalierly when noxious odors waft from unkempt quarters. Perhaps the most common surprise among visitors to Polyface is this assessment: "I can't believe it; all these animals and nothing stinks." To us, that's the highest compliment that things are as they should be.

Too many times homesteaders assume that since they don't have very many animals, sanitation isn't a big deal. I've got news for you. These animals are not potty trained. They don't flush. They go where they want then lie down in it. One animal can be as filthy as 10,000. Stay away from this pitfall.

8. Inexperienced babysitters. Invariably the homesteader family wants to go away for a couple of days. The looming question is this: "Who do we know who can watch our place while we go away?" More often than not, it's some friend who doesn't have a clue about animals or running a homestead.

The owners leave for their get-away and the fun begins. Murphy's law strikes first: anything that can go wrong will, at the worst possible moment. When does the electric fence charger quit working? Answer: when the tail lights of the owner's car disappear from the driveway.

When does the latch to the grain room get left open so the goats can enter? When does the raccoon get in the chicken coop?

When does the pig find the hole in the fence? All this happens when the babysitters come to watch the homestead.

If I solicited disaster stories surrounding this one pitfall, I'm sure I could fill several volumes. The stories are at once hilarious and tragic. Most of them you wouldn't wish on your worst enemy.

Our most dramatic story involved two very capable apprentices. They were not novices, but good grief, what are the chances? Our family decided to go to Texas for a week. Barely had we gotten airborne when a terrific wind storm came up. By the time we landed for our first connection, the apprentices called: "the cover blew off the hoop house." We had 1,000 chickens in there. It was winter. Snow was on the way.

Who could imagine such a catastrophe within minutes of leaving home? My wife Teresa says that the electric fence energizer is plugged into me. When I leave, it quits working. I wish I had an explanation for this strange phenomenon, but I don't. Perhaps it's just cosmic humor. The fact is, when you leave your homestead, things go haywire. You can prepare; you can do a check list; you can pray. When you leave, gremlins start messing with your homestead. It never fails.

The pitfall here is being flippant about the fact that things are going to go wonky. They will. That means you need to train your babysitter. You need someone who comes over for days, even weeks, to learn the ropes. You need to do some drills: "what would you do if the cows were over there instead of here where they're supposed to be?"

You might even set them up, like cutting a hole in a hose to create a leak. See if they notice it. See if they know how to fix it. Believe me, groundhogs that can't find your water hose for a year will suddenly discover it the day you leave on that coveted get-away. Don't lose your homestead and your sanity during a short get-away complements of a novice babysitter. Put in the time to make sure your babysitter can see, think, and act. Your homestead depends on it.

Chapter 4

Strategic Considerations

Planning is what makes operations go smoothly. Business gurus define this as WITB (Working In The Business) versus WOTB (Working On The Business). The easiest thing in the world is to engross ourselves in our projects; the hardest thing is to question if these projects are important.

While our homesteads may not be considered a formal business, if we don't bring some business acumen to them we'll fail miserably. We won't strategically place investments. We won't attack weak links because we'll never discover them. One of my favorite strategic planning devices is the SWOT analysis: Strengths, Weaknesses, Opportunities, and Threats. Itemizing these for your homestead can help you apply financial and labor resources to the leverage points that will give you the greatest performance.

I hope by now you realize I'm a list maker. I love lists because they help focus my energy. If my day goes wonky and I do things that weren't on my list, I put them on at the end of the day so I can cross them off. Anyone who says lists are too rigid is fooling around. Consider them simply written objectives. If you don't write down an itinerary, you'll always be late. Here is a list of strategic considerations.

1. Adequate infrastructure before acquiring animals. Purchasing animals is as easy as a phone call or click on a website, especially chicks. Accommodating your purchase of animals and metabolizing their arrival is far different than plants.

In a pinch, you can spade up a garden plot and toss in some seeds in short order. But getting everything together to handle animals is on a different magnitude entirely. I know commercial gardeners will cringe at my dismissive attitude toward putting in some carrot seeds, but I do both and the difference is profound. Carrots have a wide temperature range in which to be comfortable; chicks can only vary about 10 degrees. Carrots don't wander off. Chicks have legs. I could go on, but I think you get the picture.

Every animal has a life trajectory. Before ever acquiring the animal, think through the infrastructure necessary to accommodate the animal. Let's think about a broiler chicken, arguably the fastest life cycle on the homestead. From arrival in a box to bagged bird in the freezer: 8 weeks. Think back 8 weeks. Not that long ago, was it? Trust me, 8 weeks come and go in a blink.

Here is a checklist for broiler chicks before you place the order:

- **Brooder**--where are you going to put them when they arrive? It doesn't have to be complicated, but you need a designated place with adequate protection from the cat and the dog (and perhaps toddlers), rats, wild predators, and weather. You need a way to keep a constant temperature without a fire hazard. What kind of bedding are you going to use? Do you have enough on hand? How about grit? Waterer, feeder? Is it big enough? Remember, you'll need about 25 square feet per 100 birds up to 3 weeks. If you get 25 chicks, you need a bit more than 6 square feet.

- **Post brooder**--where are they going to grow? For sake of this discussion, I don't care if they're in a suspended cage above your car in the garage or in a portable pastured shelter. What I care about is that you have it built BEFORE you order the chicks. I wish I had a nickel for every time I've received a frenetic call: "My chicks are overcrowding the brooder box and I have to get them out but I don't have a place to put them yet." Life happens. Few things are as predictable as the fact that if you don't have your grow-out place ready before you order the chicks, some catastrophe will cramp all those days in those 3 weeks you planned to get it ready, the planned-for construction time evaporates, and suddenly you're in a crisis and everyone is blaming everyone else. Don't let it happen.

- **Inputs**--is your feed source secure? Can you store it where mice and rats won't get in it? Do you have a spare waterer in case the one you were planning to use breaks?

- **Contingencies**--are you ready for a cold snap, a heat wave? Wind? Torrential rain? Weather happens.

- **Handling**--how are you going to move the chickens from the brooder to their grow-out facility? And then from their grow-out to slaughter? You can't haul them in 5-gallon buckets.

- **Processing**--who, where, and how are you going to butcher these chickens? Remember, it's 8 weeks away. You need a way to kill, a way to scald, a way to pick feathers, a table on which to work, potable water to keep things washed down and clean, knives, chill vat, appropriate-sized bags. This is the Waterloo more times than not. If you only remember one sentence in this whole book, remember this one: DO NOT ORDER BROILER CHICKS UNTIL YOU HAVE YOUR PROCESSING SITUATION SETTLED. That means personnel, infrastructure, and final storage.

I have no intention of repeating here all the how-tos in my *Pastured Poultry Profits* book; as I said in the preface, this is a big picture book. I'm purposely staying out of the weeds because you can find the right temperature for chicks from a hundred different sources. Knowing that number and having a practical set-up before the box of chicks arrives are two extremely different things.

Now let's move to herbivores in general but cows more specifically. Of all the herbivores, cows are by far and away the easiest to control. A single strand of electric fence is more than adequate. You've probably heard the old adage about goat fencing: "If it won't hold water, it won't hold a goat." Not entirely true, of course, but goats are known as Houdinis and they have a lot of time on their hands to be mischievous.

The small herbivores like sheep and goats, when out of control, tend to be more irritating than intimidating. They're by nature less intrepid when ranging; cows have no problem confidently walking to the next county. Or running. Due to their susceptibility to predators, sheep and goats are naturally a bit more timid about going into strange territory. Not so cows.

Before bringing any four-footed beast to your tranquil homestead, make sure you have a secure place where you can hold that animal if all plans fail. The seller may tell you the animal is electric fence trained. The seller may tell you it's the most docile animal in the world. But after a trailer ride, all bets are off. Your tranquility can be shattered in less than a second.

Never let a new animal out in an electric-fenced pasture. Do I need to repeat that admonition? Never let a new animal out in an electric-fenced pasture. If the fence is not physically able to contain the animal, first put it in a corral where it can settle, get used to you, and get well trained to electric fence.

A cow is a big thing. A cow can break a board just by leaning on it. She doesn't have to be agitated or flighty to break something seemingly unbreakable. Corrals fabricated of rusty wire, rotten boards, and baler twine don't fit my definition of secure. Whatever you plan on bringing in, you need a place where that animal can exert and exhaust all its creative and brute strength talents without escaping. If you don't have such an Alcatraz, cancel the drop-off. You aren't ready.

In addition to a secure corral, you need a place to restrain the cow--a headgate. You can build the corral and the headgate yourself; no need for this to be an expensive proposition. That information is all in my book *Polyface Designs*. This infrastructure is critical because you never know when an animal might require help. A stray piece of wire wrapped around a leg can become a tourniquet with just one kick. Remember, the animal can't think through the situation; all the cow knows is that something is bothering her leg and the only thing she can do is kick at it. She doesn't have fingers to untangle it.

If you have a neighbor with a good headgate that you can use, that's fine too. The point is that if you cannot quickly restrain the animal to offer proper care, you're not ready to domicile the animal. For your own protection, physically and litigiously, and the animal's protection, you need a truly secure place to restrain your animal.

2. Matching genetics to objectives. What is your goal with these animals? Is it to have something unusual? Is it to preserve something old? Is it to feed yourself as cheaply as possible? Is it to challenge your engineering and construction ability? You might laugh at that last objective, but you'd be surprised how many times I encounter this. Of course, nobody ever writes that down as the conscious objective, but from the outside looking in, difficulty appears to be the objective.

Just like we talk about hard land and easy land, realize that there are hard animals and easy animals. To be sure, I'm not opposed to rare and minor breeds, but if you're going to push the envelope, you need to be aware of the ramifications. For example, some breeds are harder to control than others. Bantam chickens, for example, can fly over anything. They're cute and small, but they won't stay put.

They roost in the rafters over the tractor. Poop covers everything underneath the area where birds sleep. Do you want chicken poop on your tractor steering wheel, the table saw, and your shovel handles? Really? I suggest starting with an easy keeper and that is why for laying hens I prefer the old standard dual-purpose varieties. They're the most docile, the most hardy, and they can't fly very well. If you clip one wing, you can keep them pretty low to the ground.

Hybrid chickens, like the common Leghorn-Rhode Island Red crosses, are laying machines but they're more flighty and more fragile. They also have higher octane feed requirements due to their prolific egg production.

Heritage breed turkeys are similar to bantams. On our farm we raise conventional double-breasted whites because they are easy to process, easy to market, and easy to control. They don't fly; we control them with electric netting, enabling us to determine where they eat and where they poop. White feathered fowl are far easier to process to clean carcass specifications. Colored feathers add a lot of time to processing.

Goats are hard enough to control; tiny goats are even harder. The hairier and more primal a pig, the harder it is to contain in electric fence. On a homestead with children, few characteristics are as important as docility. Often the more common breeds have arrived at their numerical status because they are easy to work with. A higher percentage tend to respond affirmatively to human interaction.

Genetic memory is a wonderful thing. All plants and animals record their experience in their genes and pass on some to their progeny. This is normal adaptation to locale. Over time, certain characteristics in a region offer greater chances of success. This is why a Scottish Highlander looks completely different than a Jersey. They come from two completely different regions and slowly developed characteristics adapted to successful function in that region. Caretakers noticed certain animals thriving and others not; selecting the better ones eventually created a distinctive breed.

This is why most breeds carry geographical notations. If you're in Alabama, you don't want hairy, thick-coated Scottish Highlander genetics--unless your objective is cuteness. But realize if your objective is cuteness, it's going to create issues. In the case of a Scottish Highlander in Alabama, the issue will be extreme discomfort in July. Appetite suppression, foaming at the mouth from heat exhaustion, and overall lethargy will all plague a Scottish Highlander in Alabama in July.

Breed selection too often is based on something other than simple. While it's not flashy, it's usually the best place to start. Exotic and rare breeds generally carry exotic price tags. I'm delighted that some folks want to subsidize preserving the rare breeds, but for the run-of-the-mill practitioner just trying to hustle up your own grub, you'd do better to stick with more common breeds.

Quite a big difference exists between buying a yearling ram for $200 versus $800, especially if he's not going to be used on more than 10 ewes. The ongoing high price of rare anything needs to factor into your genetic objective, which in turn needs to factor into your overall homestead strategy. Long term strategies are all about deciding what to embrace and what to jettison. No one breed does everything perfectly; don't even look for it.

Every breed has trade-offs. I know such language is borderline offensive to folks who obsess over their particular breed, but I've tried a boatload of them over the years and have found good ones in each breed and bad ones in each breed. Our bottom line here at Polyface is that more differences exist within breeds than from breed to breed.

3. Growing at the right speed. Drinking the homestead Kool-aid is intoxicating and exhilarating. After we've read the books, made the lists, and built half a dozen homesteads in the sky, the possibilities can lead to such unbridled enthusiasm that we try to do too many things at once.

All learning is incremental and systematic, including what you need to know to have a successful homestead with happy animals. "But I want it all now!" is a harbinger of defeat. You can't have it all now; you can never have it all now.

Before he died, Dad and I sat down and made a list of potential incomes and enterprises our farm could support. That list, made in 1987, is one of my most treasured memoirs of brainstorming with Dad. At that time, we came up with 22 salaries and enterprises. We still haven't done all of them. We've morphed and certainly gone directions we couldn't have imagined back then. But I still have the list and it shows we never arrive; we just keep making incremental steps on a journey.

Homesteaders who want animals create menageries in their dreams. That's well and good, but pace yourself. Innovation is expensive due to the learning curve. Every species has a learning curve. What applies to cows doesn't apply to sheep doesn't apply to pigs doesn't apply to chickens. They all handle differently, require different infrastructure, different feeding regimens, and have different life cycles. You need time to metabolize one before adding another.

Birthing a small baby is a lot easier than birthing a big baby. At least, that's what women tell me. Be content to birth a small baby. Even small babies grow if they're properly nurtured. Your homestead menagerie will be the same way. Give yourself time to make your blunders one at a time. Too many crises in a day can leave you frustrated and depressed. This is why the average homestead cycle is six years.

That honeymoon, reassessment, and divorce pattern happens fast. In our neighborhood, I've watched some properties turn over half a dozen times in thirty years. Besides unrealistic expectations about how much work is necessary for a given amount of return, the next biggest reason for this cycle is trying to do too much too fast. Don't let that be you. One of the surest ways to escape the frustration is to bite off chewable pieces. Don't add something new until you feel comfortable with what you've been doing. To be sure, comfort can be achieved fairly quickly, especially if you spend a lot of time with the stock you have. The whole idea here is to minimize the chances of disaster by limiting novice enterprises. Get skilled in one before moving onto the next.

4. Function first, form second. I've spent a fair amount of time dealing with this idea on breed genetics, but now I want to drill down a bit more on overall infrastructure. Order and neatness do not have to be expensive.

Allan Nation used to say that profitable farms always have a threadbare look. That has certainly described ours, and still does. A common remark from visitors is how non-flashy our infrastructure is. We don't even buy fence posts. We use our own locust or cedar posts and they're not always straight, but they're cheap and functional and that's what matters.

One of our latest buildings is the carbon shed where we store wood chips and junk hay that we use for bedding. A simple pole shed, using our own locust poles, it needed partitions for the different bays for the various materials. We rotate in wood chips, for example; we don't want fresh ones drying down added to old ones ready to use. As we thought about what kind of partition walls we should use, our first inclination was to use pressure treated lumber from the hardware store. That's what people usually use in a situation like this.

But many years ago when we needed to contain our deep bedding in the hay feeding shed we couldn't afford rot-resistant milled lumber. We went to the woods and cut fairly straight Virginia pine, which is essentially a waste wood in our area (it doesn't grow big enough to mill, doesn't make good firewood, has little value) and cut poles for crude containment walls. With a chainsaw, we flattened an edge to affix them to the pole with lag screws.

These walls last about 12 years even though the pigaerator compost is against them for a couple of months each year. When they begin to rot, we simply unscrew the bolts and add the rotted poles to compost, hugelkulture beds, or throw them in a ditch. While this is not a wall that lasts for 50 years, it doesn't cost anything but some time and uses an otherwise wasted resource.

On the carbon shed, then, we duplicated this idea and it works marvelously. These crude walls give a rustic look reminiscent of corrals in John Wayne movies. These partitions will never make the front page of *Garden and Gun* magazine, but they're functional and frugal. All those cute outbuildings you see in the slick homestead magazines come with cost.

As a strategic consideration, you need to decide how much cute is worth. I like beauty as much as anyone else, but picturesque carries a price tag. To be sure, if your objective is to eventually host a nontraditional outdoor type school on your homestead, cute needs to be more important. You may want board and batten siding rather than longer-lasting, no-maintenance corrugated metal siding. This is a conversation that you and your partners need to have early on. If one person wants everything worthy of *Southern Living* pictures and the other one wants structures salvaged from the landfill, you won't have a happy homestead. It'll be in constant turmoil.

Plenty of excellent ideas never develop because somebody doesn't like the look of it. When I talk about water and mention hoisting a 250-gallon IBC tote 40 feet up in an oak tree, few things could be more functional. But a square water tote up in a tree might look funny. That's the kind of thing I'm talking about in function first, form second. If your goal is a homestead that could double as a celebrity villa, you'll be forever disappointed at the economics. Functional doesn't have to be ugly, but it certainly won't generally be the cutest option around.

5. Mutual interdependence versus cultish independence. Homesteading certainly carries a strong self-reliance and do-it-yourself (DIY) mystique. I'm thrilled to see this burgeoning interest in our culture. But often when people, including me, rebel against something, we go too far the other way. The pendulum never stops in the middle--it over-corrects.

I'm constantly asked why I don't grow the grain for our omnivores, or why I don't farrow pigs, or why I don't use draft power. Like you, I only have the capacity to develop mastery in a certain number of things. You can't be good at everything. In the end, you can't do everything. That means you have to decide what is strategically important to do and what isn't.

As an aside, parents, don't tell your kids they can do anything. Nobody can do anything. I can't possibly play basketball in the NBA because I'm too uncoordinated, too old, and too fat. Nobody can do anything. What you can do is what you ought to do. If everyone did what they ought, we'd have a different world.

The whole point of strengths-finders in the business world is that you accomplish more by leveraging your strengths than trying to overcome your weaknesses. I'm not talking about character flaws like a weakness to pornography; I'm talking about gifts and talents. In addition to personal proclivities, our homestead context (geography, climate, neighborhood) offers different options. For example, one reason we don't farrow hogs on our farm is because we have 15,000 visitors a year. Nothing is as cute as a sow nursing piglets, but nothing is more deadly than a sow nursing piglets. A protective cow won't bite you; a pig will.

In fact, she will eat you.

We buy our piggies from folks in the community who don't have visitors, thereby protecting our visitors from an aggressive sow and at the same time giving these neighbors a guaranteed above-market premium for their product. Such an arrangement benefits everyone. They don't have to accommodate visitors and we don't have to worry about some toddler getting maimed by a 500-pound sow.

In simple terms, that's how local economies work. By definition, homesteads tend toward supplying all their requirements. We talk about the self-sufficient homestead with almost cultish reverence. Sometimes we can harm homestead success by frittering away hours on highly inefficient activities. These may prove we're stubborn or perseverant, but they might actually impede overall success to our homestead venture.

"I'm not buying grain for my chickens," says the altruistic self-reliant homesteader. "I read that I can feed kitchen scraps, earthworms, grow soldier flies, and produce my own field corn." Yes, that's all possible, but what are you not doing that might be far more beneficial? Is scurrying around tending to all these enterprises keeping you from putting in as much garden as you'd like, or making apple cider, or a host of other possibilities?

The one thing that limits all of us is time. None of us has more than 24 hours in a day or more than one lifetime to invest in a legacy. Altruism and being a purist may sound good in theory, but in practice, tapping into resources and talents off-homestead for strategic skills or inputs can actually improve our chances of being more self-reliant in the most important areas.

This is why an itemized project list, from most important to dream world is strategically valuable. We can whimsically move from one interesting thing to another and before we know it, the day is gone, then the week is gone, then the month is gone, then the season is gone and we haven't made substantial progress. Figure out what is most important for you to do, for you to control. Where do you think the most critical vulnerabilities in our society exist?

Every homestead's priorities are unique to the perceptions and passions of the homesteader. Some will start with a cow. Some will start with an herb garden and home-grown apothecary. What might be right for me might be wrong for you, and vice versa. This isn't about right answers and wrong answers, or rigid roadmaps. It is about strategic considerations to pick and choose what we want to do versus what we'll patronize off-homestead. You'll never do it all, so enjoy what you can do and enjoy encouraging someone else who has different interests and skills.

Generally, people gravitate toward either animals or plants. And within the plant group, they even segregate between produce and perennials. The reason for this bifurcation in the commercial farming realm is that you can't be a master of everything. To be sure, I gravitate toward livestock, but I enjoy the garden as part of my self-sufficiency value. Could I abandon the garden and grow more chickens to pay for produce from a good local gardener? Of course, but I like the self-sufficiency of having my own vegetables. Does that make economic sense? No, but it makes heart sense and I'm not a pure capitalist.

But I would personally never grow produce to sell. It's not my first love. What I'd like is for someone to come and build a produce business here on our farm. That's the ultimate mutual interdependence.

Chapter 5

Efficiency

One of the biggest dangers in a homestead situation is the mentality that since this isn't a business, it doesn't have to make a profit and it doesn't have to be efficient. I appreciate that this book is not written to create a homestead business, but more often than not when I visit small outfits I'm struck by the laborious inefficiency of things.

Somewhere between being in a frenzied dither and being hopelessly lackadaisical lies an efficiency sweet spot. The point is that if we don't think about efficiency, we won't attain it. Every action has a technique that makes it more efficient. For example, when you're putting away eggs, pick up two at a time that are similar. Don't pick up just one egg.

The basket that received the initial nest box gathering has mixed cleans and dirties in it. As we pull the eggs out of the basket at the processing station, we pull them out two similar ones at a time. We put our dirties in a separate basket that goes through the egg washer. If the eggs are clean, we put them straight into the cartons. You can size most eggs by sight. Borderline eggs you need to weigh should be picked up together too. Our benchmark is 40 seconds per dozen, doing everything by hand. The benchmark includes no more than 10 percent dirties.

I've seen people spend 2 minutes per dozen. This is not

about working harder; it's about technique and thinking about efficiency. Every chore should be analyzed from an efficiency standpoint. That's simply farming smart. Most of us want to get more accomplished in a day; efficiency is how we make that happen. Efficiency is more than skill; it's a mindset of constant improvement. Here is my efficiency list.

1. Go loaded, come loaded. Even on a small homestead, most of our work involves toting something somewhere. We're bringing in eggs, firewood, carrying minerals to a salt box, moving the salt box, handling electric fence and stakes.

As you think about the various things you have to do, plan your routes to minimize nonproductive walk time. Ben Hartman's fabulous book *Lean Farm* drills down on this in great detail; few farming books have had a bigger impact on me in the last decade than this one. Built on the lean Toyota principles, Hartman's adaptations are revolutionary and spot on. He advocates making a movement map in order to trace your various steps in a day.

You can audit your efficiency post-day by drawing your various routes, or you can sit down with your list of the day's activities and draw it out before you start. Either is helpful to identify redundant steps. Any time you can carry stuff out and carry different stuff back you're saving time.

One of our rules here at Polyface involves carrying buckets of water to the chicken shelters. We have a portable water trough placed close by from which we dip the water we carry to the individual shelters. One of the hardest lessons to teach our stewards is to move that trough routinely to position it as close as possible to the group of shelters it's servicing. The shelters move one span (12 feet) every morning. In one week, a group of shelters moves about 90 feet; that adds up when you're carrying water buckets. In our commercial set-up, each water trough services about 24 shelters, each of which has its own five-gallon bucket perched atop. But the service benchmark for this procedure is about 45 seconds per shelter because everything is proximate.

Moving the dip trough takes a few minutes; the tendency is to put it off and just carry the water a bit farther. But if you add up the aggregate minutes, a couple of minutes moving the water trough is nothing compared to adding a dozen extra steps to a dozen water trips. Another rule is to never carry one bucket of water. Even if you only have half a bucket to fill on the last shelter, go ahead and carry two full buckets; use half of one and leave the rest for next chore time. Never, never, never carry only one bucket of water.

If you're driving a vessel out to the field, what can you put in it on the way out and put in it on the way back? One of the best illustrations of this historically is the stall-fed calf fattening protocols of ancient farming. Because fencing was expensive and rare, herding offered a quasi-rotational grazing option. Generally, cattle stayed in a corral overnight and then walked to grazing areas during the day. Fattening during all this walking exercise was difficult. As a result, stall fattening became the most desirable method to make marbled, tender beef.

Farmers would scythe forages near the fattening stall and tote it in by ox cart. Rather than go to the forage empty, they would first fill the ox cart with the manure from the stall-fed animal and take it out to the field. They could fork out the manure on the area scythed yesterday, and then fill the cart with the newly mown forage today. This procedure meant the cart never traversed stall-to-field, or field-to-stall empty.

If you're going out to get firewood, can you throw some junk hay in the cart and cover up a bare spot along the road on the way to the woods? In the next chapter we'll dig into homestead design and layout. Efficiency is why you want to put buildings and animals in a natural sequence for efficient toting. When you come in from the shop, can you gather the eggs? Or bring in some leafy greens? That way when you go out to work on that chicken waterer in the shop, you carry out the tote to put the leafy greens in. You set it aside, do your shop work, and a couple hours later when you're going back for lunch or supper, you have the tote already in the shop and you stop by the garden to grab the Swiss chard.

Whether you're walking or riding, always think about how you can use the trip to carry something. This requires thinking in circles rather than straight lines. We don't want to go out and come back in a straight line; we want to think in circuits and multiple stations or stops. Drawing this out on a piece of paper can be incredibly helpful.

2. Reduce hauling trips. Although this is related to the go loaded, come loaded concept, it's big enough for its own category. The goal here is to amass inventories to reduce routine transport of stuff.

A perfect example is chicken feed to the field shelters. Let's assume you're running a couple of chicken shelters out in the field. Perhaps they require 70 pounds of feed a day. That's quite a lot to carry by hand every day. Most homesteaders invest in an ATV or equivalent machine to get around. For the record, I'm in favor of cheap, high mileage small SUVs for this purpose. You can get a high mileage Geo or Isuzu or Subaru Forester for $1,000. Take off the doors, rip out the seats, and you have a roomy cheap vehicle that even has a roof in case you get caught in the rain.

If it blows up in a couple of years, so what? Buy another one and repeat the cycle. ATVs are specialty machines that often cost $8,000 or more. They're sophisticated, high maintenance machines. An old beater smallish SUV is generally built heavier and far cheaper to run; and it hauls more stuff. Think about it.

Back to the chicken feed. Let's assume the chickens are 300 yards away from the house. They're in a field, of course. If you haul that 70 pounds of feed out there to them every day, even if you're pushing a wheelbarrow, that's wear and tear on the pasture. And on the lane. If you didn't have to haul feed out every day, that 300 yards would offer an exhilarating early morning 5-minute hike. You could enjoy the butterflies, listen to the hawk screech, perhaps glimpse a deer grazing in your meadow before it bolted away.

But if you have to haul that 70 pounds out to the field every day, you're not going to carry it out by hand. You're going to crank up the ATV and run rubber tires and axle weight and noise out there. You won't get the exercise or the enjoyment of communing with your early morning homestead, which is one of the main reasons you decided to have a small acreage in the first place. All for a measly 70 pounds of chicken feed.

You might think I've parked on this scenario way too long, but I hope you get the point. Logistics efficiency has ramifications far beyond simple time and motion. If you have a vessel out there in the field at the shelters, then you don't have to do the daily feed haul. It can be a simple barrel or what we've used for 40 years: old metal underground fuel storage tanks. Construct a simple lid out of scavenged metal roofing and when they're empty, tip them over and roll them up the field so they are next to the chicken shelters again.

If you're getting bagged feed, you can go out in one trip and dump them all in at once so you don't have bags and string and extra work during chores. One of my favorite examples from Hartman's *Lean Farm* book is what they did with their gardening tools. His entire full-time farm is smaller than two acres. He mapped and timed his trips to the tool shed, put a time value on it, and realized that he could build half a dozen three-sided tool sheds scattered strategically around his little farm and it would save him hundreds of hours per year commuting to get tools. He purchased six of every tool and the time savings was tremendous.

It seems counterintuitive that duplication equals savings. But he was duplicating inventory to eliminate commuting or hauling tools. This is why feeding hay in a building rather than out in the field can yield such benefits. I can walk out to the hay shed and feed 100 cows in about 10 minutes without ever starting an engine or making ruts through the pasture. We make hay when it's dry, of course, when driving on the pasture is not as destructive. But during winter mud season, even walking can leave depressions.

Inventorying on location to minimize daily hauling is one of the most strategic infrastructure developments you can do to build efficiency. Before I move to the next point, let me offer one more example of this idea. If you're going to milk an animal, it's always far more efficient to take your accoutrements to the animal than the animal to the accoutrements. Think about the physics of it. If you're going to bring in the cow or goat or sheep (I don't think we're milking pigs or rabbits yet) from the pasture to a milking station, that's a lot of wear and tear and energy moving that beast and her udder of milk to a place.

Then when you're done, all that weight must move the empty udder back out to the field. And of course, the animal never walks as fast as you can, so you have to trudge along at cow pace--milk cow pace--to the pasture. Imagine if you fabricate a little milking trailer with a tarp canopy that stays in the field. You have paper towels, teat dip, and perhaps a bucket of sweet feed (a treat for her to come to the milking station easily) all stored at the little canopied trailer.

When you want to milk, you carry your milk bucket and perhaps a squirt bottle of hot soapy water out to the milking trailer. Bossy sees you coming and meets you there; she's waiting for her treat. You tie her or put her head in a simple stanchion, squirt your soapy water on her udder, wipe it with the paper towel, and set to work milking. When you're done, you release her and carry your milk treasure home as fast as you can walk. If you're concerned about heat, take a frozen bottle of water out with you and leave it in the bucket when you begin milking. That will start the cooling process in the field if you're a long way from home.

For years Dad and I milked two Guernsey cows out in the field exactly this way and it worked beautifully. We didn't have an ATV and would walk sometimes nearly half a mile out to where the cows were. And yes, we'd carry the buckets of milk home, by hand. Lots of time is wasted on homesteads bringing milking animals in to a stationary place and taking them back out. If you have to go get them, bring them in, then take them back out, that's four trips for you and two for Bossy. If you milk her in her field, that's two trips (one out, one back) for you and none for Bossy. You tell me, how soon can you pay for a field milking set-up with this kind of increase in efficiency?

3. Minimize chores. My dad used to say that no farmer can spend more than 4 hours a day on chores. Farmers need time to think, do maintenance, and make progress. While my dad's rule applied to commercial farming, homesteaders have the same principle but perhaps a different time allotment. Often homesteaders only have an hour or two a day for all their outside work due to spending most of the day in their salaried occupation. If you only have an hour a day, then only half can be spent on chores for the same reasons my dad indicated. My definition of a chore is something you have to do every day about the same time.

In other words, it's daily and time sensitive. If it's daily but not time sensitive, then it's not a chore. If it's time sensitive but not daily, then it's not a chore. Chores are the rigid have-tos of the homestead and the quickest activity to turn into drudgery. Their inflexibility makes them the first thing to be resented when spontaneous opportunities surface. "No, I can't attend the Congressman's breakfast social; I have to milk the cow."

This is why taking on an animal is such a big commitment. Using my definition, I can scarcely think of a chore involving a plant. Most plants don't need to be looked at daily at a certain time. Chores tie you down. Anything you can do to move a chore to a non-chore frees up your life.

Here's an example from our farm. The Eggmobile versus

the Millennium Feathernet. The Eggmobile is sexy and cool; everybody loves the Eggmobile. For those of you unfamiliar with it, the Eggmobile is simply a chicken house on wheels. It has a slatted floor, allowing manure to fall through and it follows the cows in their pasture rotation. The laying hens free range and scratch through the cow pats, scattering the fertilizer and sanitizing the area of fly larvae and parasites.

My first Eggmobile was 6 ft. X 8 ft. on bicycle wheels. I made some 12 ft. X 4 ft. panels out of 3/8 inch cold-rolled smooth steel rod stock and covered them with poultry netting. With six of these panels, I had 72 feet of perimeter. I fastened three of them together with simple wire loops, giving me two sets of triplicates. I'd set up my hexagon and let the birds (about 40) out for the day. At night, they'd go in the little Eggmobile and I'd move the hexagon like the leaves of a four-leaf clover. In four days, I'd push the Eggmobile up the field and re-deploy my hexagonal yard.

We could call that a homestead-scaled Eggmobile, but the birds were not truly free range. They had to stay in that hexagonal yard. One day I happened to move them across where the cows had just been and saw what they did to the cow pats. I heard angels sing and saw the stars align; it was a true epiphany. I took away the yard and birthed the Eggmobile as pasture sanitizer behind cows.

Now, of course, we run two Eggmobiles hooked together with 800 layers in them. But people wanted more eggs. For all its wonderful attributes, the Eggmobile has significant liabilities. First, the chickens spread out and go long distances from the shelter of the Eggmobile, making them more vulnerable to aerial predator attacks. Some chickens wander off in the woods and can't find their way home. We have to move it in the morning (chore) because if we move it during the day, the chickens can't find their way back to it, even if we move it no more than 50 feet.

We have to close it up at night to keep predators out. While the Eggmobile performs admirably as a pasture sanitizer, I've never advocated it as a stand-alone egg production model. Ours works because we have expansive acreage and cows. If you don't have either of those, it's a lot of trouble.

When customers wanted more eggs, we wrestled with the most efficient way to do pastured eggs without an Eggmobile. For several years, we ran 50 layers in retrofitted broiler shelters. That eliminated predation because the birds were enclosed all the time. It also eliminated having to go out and close them up at night. Lots to like, but it was a lot of chores--moving them at daybreak every single day. We were already moving the broiler shelters; this simply added more time to chores.

Michael and Joyce Plane from Allsun Farm in Australia came for a visit during that time and told me about electrified poultry net from Premier. Michael had the audacity to tell me I was "obsolete." I was offended. It took me three years to finally purchase some netting to see if it would work. I set up a 50-meter roll of the netting around one of our field shelters, propped up a corner on a five-gallon bucket so the birds could come out, and sat down to watch it not work. I knew those chickens would jump out of the netting. But in five minutes I knew Michael was right-- I should have been using this netting already. I felt ridiculous and embarrassed at my stubbornness.

At that point, we began a series of designs, starting with hoop structures, that culminated in our Millennium Feathernet, which is an A-frame on an X-truss on skids, surrounded by Premier electrified poultry net. Again, please refer to *Polyface Designs* for complete construction details and specifications. Here's the point of this saga: because this model can be moved any time during the day, the entire moving procedure came off the chore list. We could move it at 10 a.m. or 4 p.m., or any time in between. That fundamentally freed up our schedule.

Bulk feeders are a great way to reduce chores. If you have to go feed the pigs every day at the same time, the chickens every day at the same time, milk the cow at the same time, move the broilers at the same time--all of this daily routine adds up to a crunch. Anything you can do to reduce chores adds flexibility to your homestead life.

4. Get organized. Chaos is expensive. The old adage "everything in its place and a place for everything" is true. The spot that tells the tale in this regard is the homestead shop. I feel confident in saying that every homestead has a shop. It might be rudimentary, but without a vice, hand tools, power tools, rope, nails, screws and numerous other things, you won't be very self-reliant.

Fortunately, all sorts of organizing gadgets now exist to help you get neat, even if you tend to be a messy. Nobody has an excuse to have a jumble of shovels, picks, mattocks, hoes, post hole diggers and digging irons all in a heap. Wall mounts, wall hangers, hooks and other nifty contraptions now enable all of us to get organized. How many hours do homesteaders spend looking for stuff in the shop? If time is even remotely related to money, this is one of the biggest leaks on the average homestead.

Buy or make some cubbies for your assorted bolts, nuts, screws, and nails. One of the best investments I ever made was buying a dozen plastic totes. You never have enough totes, so I figured I'd start with a dozen. With these totes, every time we go do a project, we can put all the tools and hardware in the tote to keep up with them. When we move broiler shelters, we carry a fix-it tote along so we can make repairs while we're there and avoid making a special trip back out later.

Combine work stations and organize your project schedule in order to clump proximate activities. If you have to travel to the far reaches of the homestead, what else can you do while you're there? Cut that noxious bush? Move that rock? Chop some multi-flora roses? All of us can lessen running around helter skelter throughout the day. This is a never-ending refinement in efficiency because we're never completely organized. I have a two-seater ATV and I carry a heavy mattock, electric chainsaw, short-handled shovel, pliers, electric fence insulators, role of wire, can of used baler twine, sledge hammer, survey ribbon, and water fix-it tote at all times. I can fix a water leak, chop a multi-flora rose, and cut a volunteer sycamore tree growing on the pond dam all with tools I have with me.

About once a year I spend several days organizing my desk area. If you saw it this moment, you'd think a bomb went off--it's not very organized and I'm in constant angst about it. As soon as I get this rough draft done, I'm going to tackle the desk. Teresa says you can always tell what the previous month's weather has been like by looking at my desk. If it's been bad weather to keep me inside a lot, it's neat and organized. If it's been good weather to keep me out with chainsaw work and other projects, then the poor desk suffers. It's a never-ending battle to stay ahead of the paper flow. Sometimes it wins; sometimes I win.

When it's organized, I can feel the efficiency. Let's move now to the livestock on your homestead. They have to be organized too. I've been on plenty of homesteads where a daily Easter egg hunt is part of chores. Think about how many principles of efficiency are violated with a daily Easter egg hunt: maximum chore time; resources spread out rather than amalgamated; redundant movement back and forth, up and down, in and out. Surprise nests full of eggs are not funny; they represent a major leak in the homestead. Some may be rotten. Such nests represent chaos, and chaos is expensive.

How often are the animals out? If you have animals out more than once a quarter (that would be four times per year) things are not like they should be or could be. I'll drill down on some of this later in the book, but suffice it to say that statements like "there's that stupid goat again, in the garden" is not an indictment of the goat. It's an indictment of management and maintenance. Order is beautiful; chaos isn't. Order is efficient; chaos isn't.

Part of this order with livestock is not cheating. If you're running electric fence, you need to check the spark every day. Not once in awhile, but every day. Every day. Every day. I can't emphasize this enough. It's the difference between success and failure. In an area out of sight a tree limb can fall on the fence. Lightning can strike the energizer. A deer might run through the fence. Dueling calves can accidently bump into a fence and break an insulator. Check . . . it . . . .every . . . day.

Being organized requires systematic procedures. We have a six-page Standard Operating Procedure (SOP) for moving a broiler shelter. We want the feeder to be placed exactly right, the waterer at a certain height and location, the water bucket oriented with the outflow tube at 4 o'clock from bird's eye view. Push yourself to establish a routine for procedures; that will keep you from cheating.

Here at Polyface, we have miles and miles of electric fencing. We spend many hours a year trimming and cleaning these fence lines. That's part of being organized. In the end, the time spent doing that is far less than the time chasing runaway animals. In the long run, time spent getting organized always pays good dividends. If you're frustrated and depressed about the seeming chaos and dysfunction in your homestead, make a check list of changes you can make and then prioritize them. Start. You'll never regret being organized.

5. Functional access. Do your gates swing? Or like most places, do you almost get a hernia every time you open and close a gate? Well positioned and well maintained gates are a critical component of the efficiency equation. Heave-ho gates are a bummer. You don't have to install electronic-controlled automatic gates, but a good swinging gate is certainly part of efficiency.

Of course, gates cross lanes. Lanes too need to be traversable. Sometimes you'll need to drive in a field; always hug the edge; never drive willy nilly through the field. On our farm, we instruct our stewards and apprentices to view our fields like corn fields. "Only drive the way you would if this were a corn field." That's our protocol to ensure that any sod destruction that might occur is at least out on the edge of a field and not smack dab in the middle.

For general movement, you'll want a good lane. That means grading, gravel, and enough width that you can take your mobile infrastructure through without knocking off fence posts. Few things are as irritating as getting stuck in a mud hole and having to call a neighbor with more horsepower to get you out. Just one time fussing and fuming over that occurrence will more than pay for a load of gravel or some backhoe work.

Investing in good access is foundational to efficiency. You want to be able to traverse from point A to point B without feeling guilty or having to wait until the lane is passable. Mud squeezed into wagon wheel bearings and axles will bring your equipment to a grinding halt sooner rather than later. Be assured that functional access bears fruit on many branches.

My regard for efficiency is about one click short of obsessive. If you're not close to obsessive about it, you'll always wonder where all the time went and why things that seem easy to others are difficult for you. Resolve to make efficiency your middle name and you'll not regret it.

Chapter 6

Layout

Designing the homestead layout for function and efficiency is both art and science. I borrow heavily from what I've learned through the Permaculture movement ever since Bill Mollison's first Plowboy interview in *Mother Earth News* back in the 1970s. I've never forgotten my fascination with his concepts in that piece and have been trying to incorporate them into our designs ever since. His original *Permaculture, A Designer's Manual* is still one of my top five must-haves for the home library. It's expensive, but one of the best investments you can make.

Obviously the center of a homestead is the home. If an acreage already has a home on it, you don't have much liberty to adjust it. Acreage without any buildings gives you a clean slate. Lots of us would rather not complicate our homestead development with a house building project. In general, you want your home located as centrally as possible, with a nice southern exposure for passive solar gain and solariums, and in a well drained place. Building in a swamp or on low ground is problematic.

I'm also a fan of building into the side of a hill in order to have two ground-floor entrances. I describe my dream home in my book *Folks, This Ain't Normal* if you want more on that topic.

At any rate, once you have your home centerpiece established, you need to figure out where other things will go.

In permaculture parlance, visitation zones determine location. What you need to visit most often needs to be closest to the house and what you visit less often can be farther away. The way permaculture teaches it, you draw ever widening concentric circles around your house to get a feel for this visitation criteria. In general, places that need oversight the most should be located as close to the house as possible. Normally that would be something like an herb garden, or more broadly, the kitchen garden.

If you heat with wood, like many homesteads, you'll want the wood pile as close as you can get it; ideally you'd want a shed over the wood pile. But you don't want to block your view; keep that in mind. To heat our house and my mother's house next door, we have an outdoor wood furnace. This gives us central heat but all the mess is outside. Because this unit can be 50 feet away from the house, the wood pile can be farther away too.

I'm a huge believer in building simple and cheap first to see how the function fits. Routinely things change and you'll want to reconfigure. The more elaborate and expensive the wall, outbuilding or whatever, the less likely you are to ever change it even after you determine it's in the wrong place. If you want a fence around the garden, start with a simple wire or chain link or boards if you have a bandsaw mill. Once you've been there a few years and you're satisfied with the location, you can begin building your dream stone wall if you want.

We had a major shift a couple of years ago when we realized that we desperately needed a play area for children and picnic area for families. As the visitors grew and people saw our farm as a destination place, we needed more hospitality infrastructure to accommodate them. We had hemmed ourselves in with the walk-in freezer, sales building, book barn and hoop houses. Where could we put picnic tables and a children's recreation area?

No place.

Fortunately, the book barn was an outdoor storage building on skids. All we'd done to prepare for it was pour a concrete slab for it to sit on. Because the book barn was on skids, we simply removed it from the slab, broke up the slab, and in a couple of hours had a nice big area to put in some homemade teeter totters, a tire swing, some picnic tables, and corn box (yes, corn rather than sand--it's wonderful and kids play in it for hours). Had we built a real foundation permanent structure for the book barn, the re-arrangement after twenty years would have been much more difficult.

We did the same thing with our first chick brooder. When we outgrew it, we built another one, on skids, a bit farther away from the back door of the house. Eventually we built the Taj Mahal one we use now, with a concrete floor and concrete walls up a couple of feet high. But that was the third iteration. Our first mineral storage building was on four locust poles. Once we realized it was a good location and we hadn't moved it for several years, we pulled it aside and poured a concrete floor under it to keep the poles off the ground and make it harder for rats to build colonies underneath. Just a couple of years ago we finally tore it apart and, on that location, built a genuine mineral storage building.

The idea of building long-lasting permanent structures is overrated in my view. How many times do you want the same type of structure a century from now? Not very often. Normally technology, needs, and function change over a century; why not build with less investment so that when the inevitable re-structuring occurs, it's easier instead of harder? I can hear architects catching their breath. I'm not suggesting that beautiful cathedrals are improper. That's a bigger project on a community level. But on our homesteads, we need to get functional quickly and cheaply. We also need to appreciate that our needs and desires will morph with time.

Most homesteads with animals will want a weather-protected area for at least temporary shelter. The most common reason to shelter an animal is inclement weather. Cows can handle more cold wet weather conditions than goats or pigs, for sure. But no animal wants to be outside in a cold rain or blizzard. People have been building animal shelters since earliest recorded history. Shelter is a natural if not necessary component of good animal caretaking.

If the weather is bad enough that the animals want to be inside, it stands to reason that you won't want to be outside either. Locating the inclement weather animal shelter as near to your house as practical is the most sensible layout option. In my tours through Europe, I've enjoyed many house/barn combinations. Either the house is above the animals, like a second floor situation, or the animals are under the same roof line, on the same level, but separated by a wall from the people residence.

This proximate layout is common throughout Europe and I suggest it should be adopted like a new fashion statement in the U.S. What better way to take care of the animals in inclement weather than walk through a door from your house mud room and be in the barn? The objection, of course, is about noxious odors. Don't worry, we're not too far away from solving that once and for all. Animals are not stinky; only mismanagement is stinky. Hang in there; romantic smells are on the way.

If you don't want to connect the barn to your house, that's fine, but the point is, you want it as close as is practical. With that established, thinking back over the previous discussions about efficiency and logistics, you want a facility big enough to accommodate hay for herbivores and of course feed for omnivores. The animal shelter complex needs to accommodate the animals plus all their necessaries.

From a functionality perspective, realize that any time you build a structure, you create micro-climates. Here in the northern hemisphere, a building inherently casts shade on the northern side. This will naturally be cool; perhaps a place to grow mushrooms. If you have a metal side that reflects sunlight, you'll have a warm southern area. Perhaps that's where you want to put a cold frame or at least some garden beds for early production. As the sunlight bounces off the shiny metal siding, it will warm at least one temperature zone higher, if not two, out a few feet from the wall. That's a micro-climate you can leverage.

Remember too that any roof sheds water, which can either be caught in a cistern or caught in a high-moisture production scheme like mushrooms or leafy greens. Even in the middle of summer, a cool metal roof condenses dew and provides some substantial runoff early in the morning. You could place a hugelkultur bed under the eaves to catch this extra moisture. Think about how your building will affect light, temperature, and moisture; situate your other homestead components accordingly to leverage these micro-climates.

Keep in mind that wherever you house animals, you'll want to store carbon for deep bedding. In general, whatever you're going to build for animal shelter and this whole complex, make it three times bigger than you think you'll ever need. That's yet another reason to build cheaply rather than expensively. Put your money in square footage, not in expensive materials. Trust me, it's never big enough. No homestead can have too much roof space. At minimum, build it so you can easily add on in subsequent years. Think about land space and don't lock yourself in.

This homestead hub is a high impact zone, not just with foot traffic, but with heavy vehicular traffic including tractor trailers. I can imagine some of you wondering why in the world a five-acre homestead would ever receive a tractor trailer. I'll tell you: carbon (aka wood chips, bark mulch, sawdust). In our area, carbon sells for $15 a cubic yard if you get a tractor trailer load. If you get a pickup load, it sells for about $50. Before we invested in our industrial woodchipper, we would routinely buy up to ten tractor trailer loads of carbon a year; that was our fertilizer and animal sanitation investment.

It doesn't take many animals to use up a tractor trailer load of carbon in a year. A handful of cows, fifty chickens, a handful of pigs--you'll go through carbon in a hurry. And you sure don't want to pay $50 a cubic yard for it. Do yourself a favor and lay out your homestead hub to accommodate a tractor trailer.

Like the rest of the culture, farmers buy more and more stuff over the internet and have it delivered. A bulk pig feeder, for example, will not come on a UPS truck. It'll come common carrier on a pallet, either in a regular-sized tractor trailer or a pup truck, which is a bit shorter and more maneuverable. You'd be surprised how many things you'll end up getting delivered in a tractor trailer. Hoops for a hoop house, a skid of egg cartons, poultry processing equipment; the list is endless and these truckers need to be able to get into your homestead. Put down some gravel; make a turn-around area. The truckers will love you and you'll get better service.

As we think about small stock like the chick brooder or rabbitry, those need to be extremely close to the house as well. You're going to make multiple visitations per day and you don't want a five-minute walk to get there. The more compact your hub, the less time you'll spend getting to your caretaking responsibilities. We actually have five hoop houses for winter chicken and pig shelters within 100-200 feet from our house. That's lots of housed critters--space for 4,000 layers and 200 hogs--right at our back door with our retail sales building in between. Obviously we don't have noxious odors or customers wouldn't come.

On the back side of the livestock shelter you'll want a corral and load out spot. Again, you need to be able to accommodate a lengthy gooseneck trailer. That can be a long rig if the truck is a two-door cab and the trailer is 24 feet long. Give your hauler a break and make it easy for him to get his rig in and out. The reason the corral is on the side of the barn away from the house is because you don't want to take up that precious space with a corral. I can't think of a reason why a corral would ever need to be closer to the house than the shelter.

But you do need the load-out spot accessible from an all-weather surface. If you can't load in or out except in dry weather, you'll inevitably have a rain storm the night before you plan to haul stock. Then you either have to postpone the load-out or risk making ruts, getting stuck, and having a frustrating day. Adequate quarters for friends and neighbors who haul for you to be able to get their equipment in and out easily will go a long way to endearing you to their services.

While locating an equipment storage shed and farm shop adjacent to the house is wonderful if you can do it, in my experience by the time you locate the inclement weather housing, brooder, kitchen garden, and carbon storage that close, you start running out of room for other things. If you expand the hub to accommodate everything, the circle might be bigger than ideal. If something has to go farther away, I suggest the equipment shed with attached farm shop. Equipment and repairs naturally go together.

Beyond those structures will be fields, forest, and perhaps orchard. If you have hoop houses, orchard trees offer a nice edible landscape buffer around them. They provide some shade in the summer and soften the structures on the landscape. Hiding structures and presenting a kind of nested appearance creates an attractive homestead hub.

Be assured that even an extremely small-acreage homestead might have some fairly expansive structures. Imagine, for example, a 100-doe rabbitry. This enterprise could yield as much as a $50-60,000 annual income but it will require a 2,000 square foot structure. If you put chickens under the rabbits and grow earthworms from the bedding, you might raise that structure to a $100,000 per year income. As you think through where your opportunities lie, homesteads can offer some real intensive commercial options.

If you decide to jump into the value added space with a commercial kitchen, obviously that structure needs to be extremely close to your house and located off the hub. Many homesteads realize the value of direct marketing which often necessitates a retail interface. That space obviously needs to be accessible from the all-weather hub, with enough parking spaces to handle a handful of customers at a time. If you're brand new to the homesteading scheme, you may not be thinking about these kinds of commercial options. But as you lay out your hub, you want to keep your options open for future development. If you're successful, you might be surprised what you're interested in biting off five years from now.

All of this unsettledness is why you want your initial structures on skids if possible and built cheap. You have no idea what you might be changing over the next 5-10 years as you get established. Cute *Southern Living* type buildings are nice, but being frustrated for 30 years because you've invested too much in a structure to change it or move it can deter you from important innovations.

Okay, we have the hub laid out and thought through. Now let's head to the fields. Access is the first consideration when developing a homestead. If you can't get to it, it doesn't exist. Think of your hub as the heart of the homestead and the access lanes as the arms and legs. Without access you can't move animals, mow fields, install water lines, build anything, or take visitors around. You need all-weather access lanes in order to provide efficient transportation, give a sense of order, and define your landscape.

Farm lanes should be at least 16 feet wide. While this may sound like a lot of lost land, in the big scheme of things, it's not. At that width, half a mile only takes up half an acre of land. For the convenience of access, that's a small amount of land to give up. The 16 foot width is critical for maintenance and for easy animal movement. Animals aren't comfortable getting too close to electric fences. These laneways will be bounded by electric fence; no animal wants to be within a foot of an electric fence. That 16 foot corridor is essentially about a 12 foot comfort corridor. If you don't want stressed animals, give them an artery big enough to walk down without being scared of being shocked.

An all-weather lane needs to shed water and have a solid base. Depending on your region and terrain, you might be able to accomplish this with nothing more than some gravel or grading. Usually, a good lane requires a bit of both. If you have a good shale bank to dig in, you can get excellent road material from your own place.

A good excavator can grade out a lot of lane in a day. It's not as expensive as you may think. Unless you're in extremely difficult terrain, half a mile in a day is quite common. If a good dozer operator charges $150 an hour, that's $1,200 for a half mile lane. Lots of folks spend four times that much on a riding lawnmower. I'm belaboring this a bit because the most common sequence of events for a new homesteader is to run out and get animals before shelter, access, fencing, and water are in place. Then every day is a constant inefficient catch-up trying to accommodate the animals' needs with half a tool box.

At Polyface, we now lease a dozen properties in the area. We're extremely skilled at developing managed-grazing operations. But we will not go onto a place with animals until it has all-weather access, a corral and head gate, good load-out area, fencing, and everywhere-available water lines. Cheating on those requirements is like heading off on a 1,000 mile road trip with one headlight and no spare tire. You just shouldn't do it.

The most important consideration for a lane is water. I'll use road and lane interchangeably now, but I started with lane in order not to be intimidating. A farm lane is a much different road than a public road maintained by the highway department. One of the main differences is that it doesn't have to be wide enough for two vehicles to pass. But in most respects, it's similar.

Water volume and velocity destroy roads. In dry, gently rolling terrain, roads can be seen as gutters to conduct surface runoff into ponds. Advocated by Australian water guru P. A. Yeomans, this system has been perfected by Regrarians founder and design expert, Darren Doherty. The rule is 1 foot of drop per 100 feet of run; at that rate, the velocity is slow enough to not cause erosion. In this scheme, you in-slope the road so the water stays toward the uphill side. Strategic ponds along the route offer cisterns for this road water to collect.

The drawback to this design is that it makes roads far longer than they need to be. Strategically placed ponds will still catch most of the water from surface runoff without the meandering roads. If you take a lifetime of spending twice as long in time and fuel to get places, that can make some expensive water. Australia is much drier with a much gentler terrain than most of the U.S., especially east of the Mississippi River. The best design in one place is sometimes not the best design in another place.

Obviously you don't want a road going through a swampy area; situating it on a cut-in above swamp grade can get you out of the dampness and onto more solid footing. One technique we use on our farm is what I call the pond dam access. You may need to argue with your excavator about this, but the idea works. Road builders have a mindset biased toward drainage. Everything is about getting rid of water, not accumulating water.

If your lane needs to go through a little valley, the normal excavator will instruct you to put a culvert in the bottom of the valley and then dig into the access edges for enough fill to cover the culvert. I advocate a different approach. Using no more fill dirt, put the culvert in the top of the fill instead of the bottom. Essentially, you build a pond and put the culvert in the top of the dam. The dam offers a stable road surface. You don't need to move any more dirt really; all you've done is change the location of the culvert. By putting it high instead of low, you get your valley or ditch crossing but instead of draining all the water away, you get a pond on the uphill side.

The fill for the crossing, rather than coming from the adjoining slope, comes instead from the bottom of the valley on the uphill side. Rather than pushing material downhill, you're pushing material uphill onto the dam. It's always better to push dirt uphill than downhill. That means you can use more fill to great advantage so you can build up the crossing a couple of feet higher, which reduces the distance down to the crossing from the adjoining high ground; it shortens the access slope. That's a good thing too.

By placing the culvert high, you don't have to worry as much about debris washing down and plugging it up. The pond catches debris and silt, slows everything down, and protects the culvert inlet from the normal leaves and trash that flow down the bottom of a valley during a surface water runoff event. One final point on this: rather than using a 12 inch culvert, you can use two or three 6 inch sewer pipes. The pond water level comes to the bottom of the pipe. By using several 6 inch pipes rather than a 12 inch, you gain 6 inches of water in the pond. Because a pond is bowl shaped, that top 6 inches is equivalent to the next 12 inches; those top inches add a lot of water to the pond.

By using this technique every place you have a culvert, you have a small pond in addition. Even if your place is extremely small, a couple swimming-pool-sized ponds fundamentally change habitat diversity and overall functionality of your landscape. These small ponds, too, offer security in a catastrophe. If all else fails, you can always get animals water. That's the best insurance you can buy.

When we build farm roads, we like to outslope them rather than inslope them. All normal public roads are crowned and slope off both sides. That means a drainage ditch runs along the uphill side that eventually goes to a culvert to duct the water under the road and downhill. A farm road doesn't need the crown because you're not going 50 miles an hour. The whole surface can be outsloped, avoiding the building cost and maintenance of the inslope ditch and subsequent culverting.

The gentle outslope moves the water off and downhill all along the road. It's a much cheaper maintenance option than inside ditch to culvert. In addition, you want periodic diversions to make sure the water leaves the road. The two kinds are water bars and broad based dips. Any timber logging Best Management Practice manual can help you understand the construction techniques for these diversions.

In general, you want them on a 45 degree angle to the road so the flow of water does not puddle up, but rather flushes on out the upside of the diversion. If you make the diversions 90 degrees to the road, the water stops and puddles and then you eventually have to fill in the eroded upper side. These look almost like diagonal speed bumps, but they are the life of a road to keep volume and velocity from building up. Broad based dips can be placed in natural valleys to accomplish the same thing. A good road rides the hillside, like a terrace. That offers the easiest way to maintain a solid bed and to drain off water.

If you see gullies in your road, you're losing your road. It might take a few years to really notice the damage, especially if the little gullies are parallel to the road, but as the road bed washes away, you'll eventually have a notch in the landscape. What you've created is a road river. Eventually the erosion is deep enough that knocking the water off requires major excavation to create high enough diversions to get the water out onto the adjoining landscape. When we came to our farm in 1961, we had a half mile road through the property that was notched several feet. Once we paid an excavator to build some six foot high water bars, the road stabilized and it's been relatively maintenance free now for 50 years.

Grading in a well drained all-weather road is a foundational investment to a smooth running homestead. Don't delay this development project.

The final point on layout is to centralize your utilities. Part of clumping buildings that require power and/or water is to reduce electric wires and pipes. Those get expensive. Simplifying your utility interface at the hub saves money and troubleshooting if you ever have a problem.

While I'm certainly a fan of electricity, I'm not a fan of it in the barn. The statistics are compelling. About 90 percent of all barn fires are electrical. Think about it. Rats chew insulation; dust is everywhere; everything is combustible. When I read that statistic about 40 years ago in a farm magazine, I walked straight out to the shop and cut the electrical wires going to the barn. With today's LED lanterns, you can have all the light you need without running wires through the hay mow.

When we run the hay elevator during the summer, we either use a portable generator or we run heavy duty extension cords from the shop. Those extension cords do not run through the barn; they stay outside along the road. I'm a firm believer in keeping live power out of the barn.

With these basic ideas, you should be able to create functional and efficient homestead infrastructure. Over time, you'll appreciate making your early buildings portable (on skids) if possible, making things cheap enough to be torn apart and rebuilt, and giving yourself all-weather access to the whole place. Welcome home.

Wood chips drying in the carbon shed are used for livestock bedding and composting.

Cows in the hay shed stand on top of the carbonaceous diaper. This setup makes for comfortable animals without any noxious odors.

I generally don't remove stumps. Instead I shape them to minimize obstructions. After cutting the tree down, re-cut the stump as low as possible.

Next bevel the edges of the stump. (The stump will eventually be hidden in tall grass and otherwise poses a threat to equipment.)

A sculpted stump can be left to decay and is easier to drive over when converting an area into pasture.

New grass seedlings poke up through hay mulch (left over from feeding cows) in this recent forest-to-pasture conversion.

Blound: Black rounds dug into the earth make a cheap access point to buried water lines.

A block of wood on top of the blound insulates and protects it for a poor-boy water installation.

Daniel Salatin shows the proper way to hold a rabbit.

The harepen is for weaned rabbits and is moved every day to fresh grass. It's also ideal for mowing near the house.

Pigaerators stir the carbonaceous diaper, converting it from anaerobic fermentation to aerobic compost. Compost is the heart of our fertility program.

Joel demonstrates the proper way to hold a chicken.

Using air hose or other small diameter tubing to deliver water to chickens increases the capacity on a garden hose reel, offering a farther reach to a spigot. The simple valve on the tub controls flow.

The feed bin on the Eggmobile holds 2,000 pounds of feed, reducing chore time and the number of trips toting feed to the field.

PHOTO: JEFF & NOVA JEAN BUTLER

Permanent electric fence posts from native locust trees are simple, short, and cheap.

PHOTO: JEFF & NOVA JEAN BUTLER

The first flush roof water collection system is ideal for cisterns. First the runoff goes into the big pipe. Once it's full, water runs into the smaller downspout into the cistern.

The basic 10'x12'x2' chicken broiler shelter and moving dolly allows one person to move 75 chickens to fresh pasture in one minute without starting an engine.

A retrofitted chicken broiler shelter with nest boxes tucked under the end illustrates multi-purpose infrastructure standardization.

Joel's granddaughter Lauryn Salatin showcases the home-made head gate. This gate has weathered decades of use and thousands of cows.

Joel's grandson Andrew Salatin tends his flock of sheep.

PHOTO: JEFF & NOVA JEAN BUTLER

K-line irrigation ensures grass growth even in a drought.

PHOTO: JEFF & NOVA JEAN BUTLER

Grass growth after two days: note the outer sheath, where it was grazed (pruned), and the new growth, which an animal could eat but shouldn't.

Make your excavation a planned two-fer: build an access lane and a pond at the same time.

Turkey poults in our brooder utilize hovers and simple drinkers. Both technology innovations are borrowed from the industry but are scalable to a much smaller setting.

Shed sides hold chips, bedding, or compost and offer cheap building options from your woodlot.

The rafters and roof of a simple homemade pole shed are rustic and cheap, but functional.

The Raken house illustrates the stacking concept of increasing production in small spaces on deep bedding with rabbits above and chickens underneath.

Creating silvopasture using pigs as an ecological exercise massage tool to awaken the latent seed bank, creates a beautiful and functional landscape.

Pig sort boards are an indispensable tool.

A simple do-it-yourself spring tap is easy to make with roofing membrane, bulkhead fittings, and holes drilled in the intake pipe.

Joel's grandson Travis Salatin manages Khaki Campbell ducks for egg production.

Water bars are key to access road longevity. Water bars eliminate volume and velocity that would otherwise cause erosion.

Chapter 7

Water

Water is life. Without it, you have nothing. It's even more important than access, but I put access first because you'll have to get to your water. Access is still the first priority, but water is a close second.

Five sources of water exist: rivers and streams, springs, roofs, aquifers (wells), and ponds. The only two that increase net hydrology are ponds and roofs (cisterns). Everything else is a net reduction of total water in a given ecosystem.

As caretakers of not just animals but whole ecosystems, one of our goals should be to leave the landscape more hydrated than it was before we brought our intellectual and mechanical creativity to the landscape. I'm belaboring this point because the default water discussion, no matter application or location, centers around drilling a well. That's how normal people get water. I hope nobody reading this book wants to be normal. ha!

But aquifers are not limitless basins of water; around the planet, aquifer depletion is a serious issue and worsening. If we start with the premise that our interaction with the ecosystem should create more water rather than less, then one of the practical implications is that we can only do things that everyone can do. In other words, an egalitarian or democratized opportunity needs to be one of the criterions of acceptability. If everyone sticks a straw in an aquifer or river and starts sucking, at some point the aquifer or river dries up.

This is why the great Murray-Darling river in Australia, which supported steamboats a century ago, now hardly flows to the ocean. The mighty Colorado slows to a trickle. The great aquifers supporting pivot irrigation in America's west are receding at an alarming rate. To be sure, watering your five cows is a different ball game than irrigating a thousand acres of soybeans in Colorado. I recognize the difference, but it is an academic difference. Whether we deplete the ecosystem's water inventory by 100 gallons a day or 10,000, the trajectory is the same.

Let's go through these water source options first and then we'll talk about an actual distribution system.

1. Well. Obviously this is the most common and the one least likely to raise eyebrows, partly because nobody sees it. The fact that it is largely hidden creates an innocence mystique. Out of sight, out of mind.

GOOD POINTS: normally potable, ubiquitous expertise for installation. Well drillers are everywhere and the technology for pumping and installation is common everywhere. All you have to do is pick up the phone, sign a contract and pay a deposit and you're off to the races.

BAD POINTS: maintenance normally requires skill and equipment beyond your capability; you can't see what's going on down there; toxins can creep in from unknown sources at any time; it can go dry without you knowing things are critical. Plenty of folks have gotten up in the morning, turned on the water faucet, only to see muddy water and hear gurgling. Wells fail routinely and without notice, resulting in catastrophic crises.

During a severe drought in our area just a few years ago, hundreds of wells failed, resulting in a backlog for drilling that took months to clear out. Some folks had to wait a year before getting a new well drilled. I was on a farm in South Australia where in the early 1800s during initial European colonization, deeds carried a "depth to aquifer" notation. When wells were all hand dug, a difference of two meters to the aquifer had a significant bearing on property value.

This farmer who hosted me still had the original deed that noted the aquifer was only two meters deep. Then he recited the history for his property. Within a generation, that aquifer went dry and they had to go down another 10 meters to the next one. Within another 50 years, that one went dry and they had to go another 20 meters to the next one. You see where this story is going. When I visited him around 2010, the fourth aquifer at 150 meters was going dry and they hoped there would be one underneath.

The whole idea of punching holes in the earth's surface to extract water that slowly accumulated over centuries seems like deficit hydration to me. Perhaps if I saw active hydration replenishment to compensate, I'd be less opposed to wells. But I don't see organic matter being replaced (high organic matter is the sponge that holds water; it creates spaces in the soil); I don't see ponds equivalent to what the beavers and indigenous peoples created; I don't see storage as a national imperative; I see drainage as a national imperative.

In full transparency, we have a wonderful well on our farm, drilled a century ago. It's only 80 feet with a static level of 4 feet. I don't know what the gallon per minute flow is, but it's enormous. One of the reasons Dad purchased this property was because a prior owner had a daughter with cerebral palsy. Doctors encouraged the family to install a swimming pool in order for her to exercise in the low-joint-impact water. That was in the 1950s, prior to chlorine and prior to the private swimming pool as a status symbol. Domestic in-ground pools were quite novel in the 1950s.

Every week they drained and filled the swimming pool from the well in order to keep the stagnant water from turning green. I'm sure she didn't go swimming for a couple of days after filling; our well water is about 50 degrees Fahrenheit. Old timey neighbors say that during the droughts of the mid-1900s, folks would come from around the community to fill water jugs at this well. But even as wonderful as it is, we use it sparingly for the livestock and never for irrigation. We dug a pond 50 feet away that irrigates our gardens.

To be sure, I don't think drilling a well is sinful. I belabor the negatives, however, to get us to think differently rather than simply embracing the default position for water sources. We have a lot of alternatives.

2. Streams and rivers. Again, this is everyone's water; part of the commons. While I don't have any trouble using this for livestock water, I wouldn't use it for irrigation due to the straw idea we've already mentioned.

GOOD POINTS: free for you personally, even cheaper than a well. Visible, making monitoring quantity and quality a bit easier.

BAD POINTS: like wells, can become tainted with toxins without your knowledge. However, since these develop more from surface sources, water quality can generally be determined by what's going on in the watershed.

Draw-out can be ticklish due to dramatic changes in flow. The lowest summer level is quite different than flood stage. Whatever pumping set-up you have, including the point of suction, needs to be able to work at both low flow and high flow. Plenty of stream-side installations have been wiped out in a flood. I recommend building a pump house up on a pedestal of fence posts in order to make sure you're safe from flood waters. If you have a high bank on one side, of course, that can serve as a pump staging area.

Stream water is often turbid, which seems to affect livestock more than organic waste like bacteria or small amounts of fish manure. A sand filter can be worthwhile in a stream set-up, especially if it's a slow-moving stream. Fast-moving streams tend to clean themselves with high oxygenation and natural flushing due to cascading turbulence.

If your stream or river grows duck weed and has lots of salamanders and frogs in it, you can be sure it's safe for your livestock. If you can't find any of these critters along the edges, you probably should pull a sample and see what's wrong. If water won't support life, it's not a good source for your animals.

If you want to direct water your animals from a live source like this, always constrict the access. Animals do not feel comfortable if the distance between electric fences is 8 feet or less. On one of our rental properties we use a fast-flowing stream for live water and have numerous points of access that are simply a V. The wide part of the V is 8 feet on the access side and it goes to a point on the other side of the stream. Because the stream notches we hang a weight (piece of metal, rock, wood) so the V conforms to the depression of the stream and the cows can't get out underneath it.

We call these a trapeze because often we use a junk T-post as the weight and it looks like the bar on a trapeze. We use two kinds of techniques to do this. One is a separate piece of wire for the trapeze so everything dangles underneath the main wire, which goes straight across the river at bank height. The other technique is to use only one wire and affix it to either end of the trapeze. If you use a rock, it won't look like a trapeze; it'll look more like a V. Depending on the creek's topography, you might get by with a simple weight like a rock, but often you'll want more horizontal length above the creek, like a T-post. If you use the one wire system, and especially if you use a rock as a weight, make sure you wrap your wires well, whether you cut the wire or twist it, to have spark continuity.

In either case, the whole thing swings, of course, if a flood comes and reaches the weight. If flood debris breaks the wire, it just swings around parallel with the creek and part of our flood clean-up aftermath is to swing these broken ends back around and attach them. Pretty simple and this set-up works like a charm. Because it's so restricted, the cows don't loiter in the stream. They step to the edge, drink, then back out.

3. Springs. If your property is blessed with a spring, be thankful. Flow will generally fluctuate dramatically depending on precipitation rates. Always calculate flow at a dry time because that's the rate you can count on.

GOOD POINTS: usually potable; cheap to access; visible.

BAD POINTS: can become tainted due to something you can't see; can sometimes deprive someone else dependent on that water if you capture it all; often fluctuates dramatically depending on weather.

I won't go into all the fine points of capturing a spring; plenty of information exists for those mechanics. What I think I can add to the discussion is an appreciation for possibility and a poor-boy way to tap. A one gallon per minute flow rate looks like nothing, but it adds up. One gallon per minute is 60 gallons per hour and 1,440 gallons per day.

If your homestead animal menagerie requires less than 1,000 gallons per day, a trickle of water will supply it. For fun, I suggest you take a one gallon vessel and put it under your kitchen sink faucet. Watching a clock, adjust the flow until it's coming out at one gallon per minute. You'll be astounded because it ain't much. You'll say "You mean that little trickle adds up to 1,440 gallons per day?" Yes.

In general, animals will drink one gallon per 50 pounds of body weight per day. Each species drinks a different amount, and of course it fluctuates dramatically depending on weather. This past winter we had snow and sleet and did not give the cow herd any water at all; they just picked it up from stockpiled forage as they ate. When I returned the water trough to them, they wouldn't even drink, so I know they had plenty of water. You'll have times of low consumption and high consumption, but for ballpark thinking, imagine one gallon per 50 pounds of live weight.

Metabolism plays a part. For example, 50 pounds of broiler chickens on a hot day will drink about 1.5 gallons of water while a 1,000 pound cow will only drink about 12 gallons. The fast-growing, high metabolism broilers drink way more water per pound than the slower growing more lethargic cow. These are ballpark figures only, but I'm using them to encourage you to tap into that little trickle of a spring, to appreciate that even a trickle can add up to substantial amounts.

Okay, we've established that even a tiny spring can probably water your homestead livestock operation. Now, how do you tap it? You have to capture it. Again, all sorts of elaborate systems exist if you want dependable potable water for your house. But for livestock, open tapping is fine. You don't need to put a cover over the spring to keep the deer away. All you need to do is capture the flow.

I've done several of these installations over the years and they work beautifully. The set-up I'm describing will work for any spring up to about 30 gallons per minute flow. For you novices, you can't get 30 gallons a minute out of your kitchen faucet. About the most you can get out of that is 5 gallons a minute. At 30 gallons a minute, you're getting 43,200 gallons per day. Just sayin'.

Take a piece of metal roofing at least 24 inches wide; 30 inches is fine as well, and cut it about 6 feet in length. A spring will always flow into some sort of channel. Unless it's a huge spring, that channel will be something you can step across. The actual water flow surface might be a foot across as it spreads out.

From the plumbing supply store, get a one inch bulkhead fitting and a couple of two inch bulkhead fittings. Install the smaller bulkhead fitting about 8 inches above one side of the metal roofing. Install the two bigger ones about 8 inches from the other side. You've just made an impermeable diaphragm with three outlet ports. The smaller port will be under the water and the two top ones will be for overflow. You can make a simple screened intake for the spring side of the diaphragm by drilling 1/8 inch holes in a one foot piece of plastic pipe with a stopper on the end. That stopper can be a true barbed end stopper or you can whittle a round piece of wood and jam it in for a plug. That's what I do. Frugal to a fault.

Take your piece of roofing (without the pipes installed yet), a pick and a shovel up to your spring. Pick a spot a few feet away from its source (you never want to disturb that actual spring source) and begin digging a trough at 90 degrees to the flow channel. Dig it long enough to accommodate your roofing diaphragm. Go about a foot below grade. Insert the piece of roofing in the channel; you'll have to play with this a bit to get a decent fit. Once you're satisfied that it's fairly level and not sitting on a high rock tip or something, start filling your channel with the cleanest dirt available. You want material as impervious as possible because this is your poor-boy seal around the bottom of your dam diaphragm.

Once you have your channel filled to previous flow channel grade, place some rocks against the downhill side of the diaphragm to stabilize it and begin digging out a bit on the uphill side (the spring side), gently scooping the material against the diaphragm. Add material to each side a bit at a time to not overbear one side as you heighten the berms on your flimsy sheet metal dam. Install the intake pipe on the bulkhead fitting on the inside. Install the outflow pipe on the bulkhead fitting. Keep filling both sides of the sheet metal diaphragm and even around the edges until you know that the basin will flow through your top bulkhead fittings. Attach your outlet pipe to the bottom bulkhead fitting and your 2 inch pipes to the top two fittings for your overflow; congratulations, you're up and running. For a photo of this setup, see page 97.

Now watch the basin in front of your diaphragm fill. Once it covers your inlet pipe, it'll start flowing down the 1 inch pipe and you can take it wherever you want. Few things are as magical and valuable as water in a pipe. This is a simple poor-boy way to harness a spring for livestock water.

The sheet of roofing is the impermeable membrane that will trap the water. The lower fitting captures the water and the two upper ones give overflow capacity. Behind this simple dam you'll have a pool of water. The set-up I'm describing is to start a gravity system. If your spring is not on a high point, which is extremely common, you might need a combination gravity and pumping arrangement. If you can gravity it closer to where you have power, you can run it into a cistern and pump it out to your fields.

If the spring is extremely low on your terrain, you may need to pump straight from the spring, in which case you can omit the bottom bulkhead fitting and just install the two top ones for overflow. Your pump will suck water out of the pond created behind your diaphragm. In even extremely small springs, you'll be amazed how quickly your little bathtub-sized basin fills with water.

The seal at the bottom of your diaphragm doesn't have to be perfect. This is not a high pressure system. Bits of silt and debris will quickly help plug any cracks at the bottom of the roofing panel. Water, which always seeks the easy route, will quickly find your bottom bulkhead fitting and flow out. If you fail to capture every single drop of spring water, don't fret. All you need is enough. Over time, your basin will collect soupy muck that you can plaster around your diaphragm for a better seal.

4. Ponds. I'm a huge fan of ponds because they offer multiple benefits for the ecosystem. They also put us in the mindset and activity of the beavers who created an 8 percent water coverage of North America before the Europeans came. Today, it has been reduced to 1 percent; that's a lot of water we no longer have. Some 200 million beavers inhabited North America back in the day. Based on skeletal findings, some were nearly as big as a small automobile. Imagine the teeth on that dude.

GOOD POINTS: visible inventory; increases the commons; reduces flooding; cheaper than wells; creates riparian habitat; offers unique microclimates due to thermal mass; grows fish and other aquatic life; swimming; boating; soft water.

BAD POINTS: not potable but plenty good enough for animals; can freeze in winter; must be engineered to accommodate massive surface run-off events; can attract bears as a playground and bears sometimes enjoy messing with your intake installation.

In this book I don't need to go into the details of how to build a pond; plenty of earthen pond books exist to provide you information. What I will do is explain some nuances you might not find in those pond building books. The number you need to know is this: about 30,000 gallons. That's one acre-inch of water. In other words, if you covered one acre with one inch of water, it would be 30,000 gallons. It's actually exactly 27,225 gallons, but that's hard to figure in your head, so I just round it to 30,000 and it's close enough for ball parking. I hope that's fair enough and you won't accuse me of cheating.

Small acreage owners typically look the other way when I start talking about ponds because they think they don't have enough land to have enough surface runoff to fill a pond. But like the spring trickle, I want to encourage you with possibility. One acre is not much land. If you live in a 30-inch rainfall area, which includes everything east of the Mississippi River and a lot west of it, in a typical year you'll have around a third of that water run off. In other words, sometimes the soil is already saturated and can't hold anymore. Sometimes a torrent comes too fast for the soil to absorb.

A rule of thumb is that at least one-third of the precipitation that falls on a given square yard of land runs off the surface. That includes everything from desert to tropics, mountains to valleys. In a 30-inch rainfall area, then, one-third of that would be 10 inches per year. Remember that number I gave you a few minutes ago? 30,000. That's just 1 inch of runoff per acre, in gallons. What are 10 inches of runoff? 300,000 gallons. A cubic foot holds 7.5 gallons of water.

A 300,000 gallon pond would need 40,000 cubic feet of space. What does that look like? For reference, imagine an 8 ft. X 7.5 ft. X 45 ft. long tractor trailer body. That's 2,700 cubic feet. So a pond capable of holding 300,000 gallons would be equivalent to about 15 tractor trailers. Well, what does that actually look like in the field? Imagine a pond 100 feet in diameter averaging 5 feet deep. That's it. What do you think about that? Go ahead, just think and imagine it for a moment.

In a 30-inch rainfall area, that's what you can fill in an average year from a one acre watershed. Folks, the amount of water leaving our landscapes boggles the mind. We have more water than we can ever imagine. Think about an arid 10-inch rainfall area; that's 3 inches of runoff per year. Imagine 10 acres of watershed and you're back to the same numbers. In arid country, 10 acres is nothing. The truth is we have not begun to leverage our resources, especially in regards to water.

Now, what about cost? Again, as a rule of thumb, excavation will cost about a penny a gallon. Some places it'll be more and others less, but that's a ballpark figure. That means our 300,000 gallon pond will cost about $3,000. What does drilling a well cost? $7,000? $8,000? $10,000? The beauty of a pond is that we're actually increasing the hydrology of the landscape and we can walk out any morning and see how much water we have. We won't suddenly get caught with a dry well.

I hope this little exercise convinces thousands of small acreage folks to call their local excavator and build a little pond. Don't call the government and ask for matching money. Trust me, by the time the government engineers get done with your little project it'll cost way more; your portion of the cost share will be more than if you just did the thing yourself. Rely on the wisdom of experienced excavators in your area. Vet them by asking to go see a couple of ponds they've built over the years. Are they functioning?

The best water in a pond is 16 inches under the surface. If you draw the water off the bottom it'll be funky and dark. If you pull it off the surface, it'll be full of fly legs and bird poop. The best water is 16 inches under the surface. We attach an intake to a flexible non-collapsible suction pipe, affixing a piece of metal and a float to it. The float is an empty plastic fuel can. With the float and the piece of metal, the intake can be positioned at 16 inches below the surface no matter how high or low the pond level.

When you put the PVC pipe through the bottom of the dam, that slick plastic creates a weakness that can allow water to seep alongside it. Over time, that seep will eat away at surrounding material until you have a leak. You can buy pond collars or you can make poor-boy ones out of some concrete to stop this seepage from beginning. The idea is to put in a couple or three obstacles along that PVC pipe so that when water comes down alongside it, the water has to squeeze out to get around.

You can run everything off a siphon, but I always hate to run my whole water system off a siphon. And if the pond gets low enough that the intake must suck from several feet, a siphon can become weak. I like water to come out under pressure, pushing rather than pulling; hence, the pipe in the bottom of the dam rather than a siphon out the top.

Now we need to talk about low pressure. In both the spring and the pond set-up, if we're going to use gravity, we're stuck with extremely low pressure. Another important number to remember is .432. That's the pounds of water pressure per vertical foot. In other words, a column of water (or a pond) 10 feet high will create 4.32 pounds of pressure. That's not a lot.

Normally in gravity systems you don't have the privilege of heading immediately over a cliff. The pipe, whether buried or running on the surface, will head slightly downhill with the terrain. That means you might have to go 100 feet or more behind the spring diaphragm or the pond dam to get 10 feet of head pressure. If a slight flattening or elevating occurs, that pressure is too low to push out an air bubble and you have vapor lock. On all our gravity systems, we place a T in the line about 100 feet below the intake to bleed off air bubbles in that initial low pressure period.

Once the pressure builds a bit, it'll blow out bubbles and you'll be fine. If you've never dealt with low pressure systems, this can be aggravating. But once you learn to be patient and let the water seek its level--sometimes you have to walk away from it overnight and let it gurgle to fill--you'll be comfortable with it. Few systems are as wonderful as gravity-fed water distribution. On our farm now, we have nearly 10 miles of water lines, all gravity fed from ponds we've built permaculture-style in high ravines. Due to the elevation, we have 70 pounds of pressure, which is nearly double a regular domestic system. It's like a firehose and as long as gravity works, we have water. It's just short of miraculous.

To be sure, on a small acreage you may not have enough elevation to run a complete gravity system. Sometimes a hybrid is in order. Like the spring example, you may need to gravity the water from the pond to a pump that then sends it back out under pressure. On a couple of the properties we lease, we've done that. We gravity the water to a 1,200 gallon cistern with a pump situated on top. The pump is located as near as practical to the farm hub where we have dependable power.

The pump sends the water back up to the fields, some of which are higher than the initial pond. Now let's think about another number: 2.3. That's how much pressure it takes to send water up a foot. In other words, if you have 40 feet in elevation, you'll need 92 pounds of pressure. As a rule of thumb, regular off-the-shelf garden hose and plumbing fittings are good to about 80 pounds of pressure. More than that and you'll start blowing

O-rings and valves. If you need to push water higher, you have a couple of options.

One is to get a high pressure pump. That is expensive and sophisticated. And you'll have to get expensive fittings to go with it. The second option is what we've done: install two pumps. We lease one property where we pull from a spring that is absolutely on the lowest square yard on the whole property. Fortunately, right behind the spring is a steep bluff. I mean a steep bluff you can hardly climb. We have a pump at the spring that sends the water up the bluff. At the top of the bluff, we have a barrel and another pump. The first pump fills the barrel and the second pump pulls out of that barrel.

Be sure to use a low-flow protection switch on any pump. It'll save you consternation and lots of money. You never want to run a water pump dry. The low-flow protection switch turns off the power to the pump if the water ever gets below the suction hose. Because electricity doesn't care whether it goes uphill or downhill, we run a cord from the property owner's house located at the base of the bluff. It's only 100 feet up to the top of the bluff (about 30 feet in elevation--I told you it was steep). That second pump sends the water on back half a mile to the far end of the property. By using two pumps, we can run everything on 70 psi which lets us use common pipe and fittings.

What if your land is flat? Here is another big concept: the only way to have 24/7 pressure without energy is gravity. Pressure tanks use energy to keep the pressure up, like a coiled spring. A lot of latent energy gets used in just keeping that spring coiled. The problem is we don't use water consistently. We need a bunch to fill a bucket or take a shower, then we don't need any for awhile. Water use comes in fits and spurts.

Making sure the water delivery system can satisfy the sudden use burst requires a lot of hardware and energy on the front end--unless we can maintain the pressure with gravity. That is why cities use water towers. Water can dribble in all night and then offer the burst when everyone gets up in the morning and takes a shower. It's always cheaper to dribble water to a high point and let it gravity feed back than it is to maintain enough power all the time to meet the demands of a use burst. Gravity maintains flow and pressure in real time without costing anything.

How can you use this principle on a homestead? Imagine you have a big oak tree. You could pull a 250-gallon IBC tote 40 feet up in that tree and dribble water into it all day. When the five cows come to drink from their trough, you've got a ballast waiting to deliver it at pressure in real time without cost. Goodness, you could rig up a pump on a bicycle and fill that tote once a day. At 40 feet you'll have 17 pounds of pressure (40 ft. X .423 pounds) coming through your system. If you're not sending the water too far, that's enough pressure to satisfy your needs.

As soon as water begins moving through a pipe, it creates friction against the sides. At any given pressure, you can increase flow by increasing pipe diameter. As a rule of thumb, from 1/2 inch to 1 1/2 inch, the flow doubles every quarter inch increase in diameter. In other words, a 3/4 inch pipe, at a given pressure, carries twice as much water as a 1/2 inch. A 1 1/4 inch pipe carries twice as much as a 1 inch pipe. Sometimes it's cheaper to increase your pipe diameter than it is to increase your pressure.

One permaculture concept is called a turkey nest pond. This is simply an excavated pond on top of a hill that acts as a cistern. You can dribble water into it, or fill it occasionally, and then gravity feed from there. I have a friend in Australia who uses this concept on his farm. He has a pond located strategically to catch run-off, but it's not high enough to gravity feed over the farm. He has a hilltop not far away and built a turkey nest pond in the top of it. Once every other week he fires up a gasoline water pump and by the time it runs out of gas, it fills the pond. That gives him gravity water for two weeks.

I hope by now you're at least intrigued enough with ponds and gravity systems to play with them. You can use GPS altitude applications on your smart phone (yours, not mine; I don't have one yet) or a topographical map, to play with some options. Don't do the first thing that comes to mind. With all these possibilities, take the time to figure out which one is best for you. For extremely accurate terrain measurements, you can use a piece of tubing, filled with water, and a pressure gauge on one end. With a partner, walk your water line idea and record the pressure.

I had a wonderful engineer friend who taught me this when I had the idea of trying to siphon water from a spring that was on the wrong side of a ridge. I wanted to siphon it to the other side but didn't know if it was too high to suck. You can't siphon more than 32 feet or you pull absolute vacuum. He had a 100 foot piece of clear tubing. We unrolled it and took pressure readings every 100 feet (remember, .432 pounds per vertical foot) and found out it was 31 feet. This is a way to get an extremely accurate elevation reading over a distance. I did not install the siphon, but I did have the idea that with a backhoe, we could dig a 10 foot deep trench through the ridge and reduce the height of the siphon. That's still in the back of my mind.

5. Roofs. I hope your head is spinning enthusiastically with possibilities because I'm going to leverage the spinning with some more cool stuff.

GOOD POINTS: controlled source; soft water; adds to ecosystem water inventory; reduces flooding; potable.

BAD POINTS: cost of cistern; limited in extremely dry areas.

Are you ready for another number? .625 gallon. That's the amount of water in 1 inch of depth on a square foot. In other words a 1 inch rain will generate .625 gallon per square foot of impermeable surface. Isn't that a cool number?

Let's play with it a bit. Imagine a 1,000 square foot roof, which isn't that big. That would be an extremely modest house or barn. A 1 inch rain will generate 625 gallons of water from that little roof. Well imagine if you live in a 30-inch rainfall area, like we talked about above. That would yield 18,750 gallons of water per year.

Just for fun, let's think about what some animals would drink. A herd of 5 cows would drink 4,500 gallons per year. Let's throw in 100 laying hens (assume you're selling a few eggs). That's another 1, 100 gallons per year. How about 1,000 broilers? That's another 3,000 gallons. What else? How about 10 hogs? That's another 3,600 gallons. Are you tracking me here? We've got a homestead with 5 cows, 100 layers, 1,000 broilers, and 10 hogs and we've used a total of 12,200 gallons of water. We're still a long way from 18,750 gallons.

And that 18,750 gallons is just from the house. If we include the outbuildings we'll probably triple it to 56,250 gallons--let's just say 50,000 gallons for fun. That will water a lot, a lot, a lot of animals. Because well drilling is by far and away the most common water acquisition method in the U.S., we don't have a vibrant cistern market. But in Australia, where literally every house, both urban and rural, has a cistern attached to it, the vibrant trade creates a price per gallon about a quarter what it is in America.

I'd love to see that changed. Unfortunately, the rule of thumb in America is about 80 cents per gallon of storage. To be fair, that is potable storage, so we're talking about food-grade plastic cisterns. On our farm, we have five 3,000-gallon potable water cisterns buried at the apprentice cottage that catches roof runoff. That kept us from drilling a well, gives us soft water (well water here is hard due to the minerals), and offers a great ballast. We can unscrew the manhole and look in any time to see how much water we have.

We have a simple first flush on it. If you're not familiar with that terminology, it's simply a PVC pipe adjacent to the downspout from the guttering. The water comes off the gutter and fills that tube first (about 20 gallons) and then flows over the top of it into the pipe that fills the cistern. That way the initial bird poop, fly legs and leaves that inevitably collect on a roof get routed into that initial tube and the subsequent cleaner water goes into the cistern. The first flush pipe has a pinhole in the bottom that lets it drain out over a day or two so it's empty and ready to receive the first flush when the next rain comes. For potable water, you should also install a good carbon filter for all the water that goes into the cistern. For livestock only, that's not necessary.

In the old days, every barn in our area had a large cistern, 20,000 gallons, alongside it because wells were too difficult to dig in our limestone karst geology. Nobody in our area had wells until after 1900. Livestock drank from springs, rivers, ponds, and from tanks supplied by cisterns. The most common weakness of cisterns is they're too small. The goal is to never run it over. If your cistern is big enough to hold all the water from the roof in one rainstorm, you're in good shape. Of course, it doesn't have to hold all the water generated in a year at once because you'll be using it between rain events.

Cisterns are normally buried or are at least at ground level. In homestead design, if you can locate it on high ground, you might be able to gravity feed to your fields from the cistern. Normally cisterns need pumps because they're low in the terrain. If potable water is not important, for livestock you could simply excavate a swimming pool sized hole and line it with pond liner or plastic; that would be much cheaper than buying a container cistern or building block walls.

Again, lots of material is out there about cisterns and I don't want to clog this idea book up with the minutiae. I hope this chapter has piqued your interest in water sources other than wells, and especially in water sources that add to rather than depleting the commons.

Chapter 8

Sanitation

Few issues are as problematic on homesteads as livestock sanitation, probably due to cramped spaces both inside and outside. Most homesteaders think their limited area makes overgrazing and dirt barnyards okay. After all, it's not a great big field. It's not a great big dirt area. It's not a mountain of manure. We're tempted to dismiss small insults.

They aren't okay. Most lawns receive much better care than homestead pastures. Especially if you have a horse. Probably the worst land stewardship, on average, occurs in horse paddocks on small operations. Be warned; I'm going to set a high bar in this chapter. The reason for this high bar is because I want you to be as ignorant as I am about diseases and sickness.

Most livestock books spend a great deal of time discussing various diseases. I won't. In my experience, they're a footnote to your life if you take care of things up front. The dirty, unhappy animal is the one most prone to sicknesses. In addition, weaknesses are highly heritable. Refusing to mask problems with vaccines coupled with ruthless culling will ensure your menagerie stays relatively trouble free. The other component is habitat, and that's what I want to drill down on in this chapter.

In my experience, most domestic livestock sicknesses are caused by unsanitary conditions. Putting attention on that takes care of the lion's share of problems.

Nature sanitizes two ways: rest and sunshine or vibrant decomposition. A third related way is through diversity and we'll deal with that at the end of the chapter.

Rest and sunshine. The key in an outdoor setting is to vacate the area long enough and often enough to create host-free periods flooded with ultraviolet light from the sun. Pathogens require hosts. If they're denied hosts, they can't propagate. Nearly all pathogens are species-specific. Even species you'd think would share pathogens often don't, like sheep and cows.

The rule of thumb is 21 days host-free at least twice a year is enough to break the virulence of pathogens. That doesn't mean they won't exist any more; it simply means they aren't at sufficient strength to be a problem. Eradication is never a practical or even a desirable objective.

Any outdoor animal hosting situation on which the same animal sets foot every day will eventually develop a parasite and pathogen load. Denying access to a host is critical to reduce infestations. Realize that these pathogens are generally microscopic; that means they can't travel very far. Eradication is neither possible nor realistic.

If you have two cows on a 3-acre field, moving them every couple of days to a new spot, even if it's adjacent to where they were the previous couple of days, is enough to break the pathogen cycle. These microscopic bad guys need easy access to mouths, hoofs, and feet. We'll go into the logistics of these subdivisions in another chapter, but for now, I want you to appreciate the rationale and the effectiveness behind rest and sunshine. Old-timers talk about the benefits of "new ground." Normally given in the context of growing crops, it's just as important in livestock management.

By denying access to a piece of pasture for a couple of months, you've created "new ground" through management. You don't have to go searching for it in virgin wilderness. You can build it into your management program with strategic animal placement.

Almost all animal quarantine protocols, as well as the incubation period for most diseases, is 21 days. In other words, if you buy an animal and it doesn't get sick for 21 days, chances are it's healthy and won't bring a communicable problem to the rest of your animals. A good preventive practice is to keep new animals separate for 21 days. If you can't, you can't, but it's a nice practice if you can.

That also means that you don't want anything on the same ground or accessing the same spot more than 21 days at a time. From initial contamination to final expression, 21 days is a commonly accepted time frame. In other words, it takes 21 days for a problem to incubate; if a problem has not expressed itself in 21 days, it probably won't; if you have a problem, 21 days of exclusion will drop virulence low enough to be controlled by natural immunity.

To make sure you don't miss the ramifications of this principle, think about a dog kennel. A pig pen. A goat milk stand. Anything where the same type of animal, or especially the same animal, accesses every day has a high likelihood of developing pathogen virility. A path traveled every day. A loafing yard accessed every day. The ramifications of this principle are profound.

Managing our livestock with the 21-day exposure and exclusion concept in mind means lots of movement. That inherently drives water, fencing, and shelter infrastructure. For the record, I've never understood the allure of a barnyard. Homesteaders too often romanticize a barnyard. To me, a barnyard is a stinky, pathogen-laden cess-pool of shameful animal husbandry. That's a strong statement, but lots of vet bills have piled up in a barnyard.

You want your animals out on new ground, away from yesterday's manure and urine and the accumulating pathogens.

Loafing and grazing ground needs rest to create a "new ground" option. And while the animals are gone from the area, the sunshine bathes it in cleansing ultraviolet light. This is why migratory patterns in wild herbivores works. Facilitated by seasons, predators, and fire, this movement protected wild herds and flocks (birds) from being overcome with pathogens.

This movement duplication and its protective ability do not change with scale. Whether you have two animals or 2,000, the constant access of pathogens to a host and the constant ability of pathogens to proliferate doesn't change. The principle is the same. Just because we have animals on a small scale does not exempt us from the pathogen principle and the 21-day protocol.

To be sure, sometimes you don't want animals outside. Weather conditions are not always friendly. While animals are tougher than most humans realize, most of us don't just want our animals to survive; we want them to be comfortable. Now that we've examined the outdoor principle of rest and sunshine, let's examine an indoor housing model, which necessitates understanding the second way nature sanitizes.

Vibrant decomposition. Well-made compost piles are certainly the most common example of this sanitation principle. All sorts of things, including toxins, break down in compost piles, from antibiotics to pathogens.

While compost certainly uses natural processes for decomposition, a compost pile is completely unnatural. Nature uses these fungi, bacteria, and nematodes to break things down, but at a much slower rate. The man-made compost pile, beautiful as it is, only occurs through manipulation. Only with human intervention does that much carbon, nitrogen, moisture, air and microbes all get together in the right ratio at the right volume.

As anyone who has ever built a compost pile can attest, it has a definite set of protocols. Violate any one of them, and you don't have a successful compost pile. A successful pile heats, but not too much, and the finished product smells earthy and rich, not fermented or ashy. Sir Albert Howard, godfather of the organic farming movement, codified the metes and bounds of successful composting in his iconic 1940s book, *An Agricultural Testament*. It's still on my top 5 must read list.

The individual components are these: carbon, nitrogen, water, oxygen, and microbes. If these are in the right ratios, the compost pile works; if they aren't, it won't. Elegantly simple. These elements describe the compost pile components.

Two other protocols are also essential: ambient temperature and mass. You can't build a compost pile when everything is freezing and expect it to start. The ambient temperature needs to be well above freezing in order for the microbes to come alive and do their thing. As long as you build the pile on a 50 degree F day and it has a couple of days to get going, it will perform nicely through almost any ambient temperature. As long as the core is functioning, cold on the outside won't shut it down.

Now to mass. Minimal size is about 3 ft. X 3 ft. X 3 ft. You can't build a compost pile 12 in. X 12 in. X 12 in. It won't work. When I explain this to kids, I tell them the tiny pile won't work because the community of microbes needs enough room to build houses and schools and libraries. Kids get pretty wide-eyed at the notion of bacteria reading a book. The point is that maintaining an active microbial community requires enough core mass in to ensure the relationships and negotiations required. One bacteria trades a mineral with another bacteria. A compost pile is literally a commercial district of trade.

It's also a war zone. Everything is eating and being eaten. One microbe attacks and gobbles up another; the attacking one is in turn attacked and gobbled up by another. Welcome to the world of decomposition. All this activity requires moisture, just like we need to stay hydrated when we're exercising and sweating. This activity literally uses up water within the pile. And just like us when we're exercising, the activity generates heat. The heat cooks and neutralizes many pathogens, just like a fever cooks diseases in our bodies. Fever is one of our defensive mechanisms actuated by our immune system.

I'm belaboring the activity within the compost pile and the necessary mass to drive home the point that dimensions matter. To get the benefits of microbial activity, you need not just height and width; you also need depth. You could have a 5,000 cubic foot compost pile, but if it's only 6 inches deep, it won't work. A compost pile requires three-dimensional cubic mass. If all the parameters are in place, you have a decomposition war zone where the good bugs destroy the bad bugs. It's not sterile. What you do have is vibrant decomposition.

We've talked about filth over-riding the immune system. That can certainly happen. But the opposite is not acceptable either. Ecology, whether on our skin or in our gut or in our yard, is never about sterility. It is all about providing a habitat in which the good bugs beat the bad bugs. About 95 percent of all microbes are beneficial; only a handful are destructive. That is why if we provide a habitat conducive to the good ones, they'll overwhelm the bad ones. The secret is not concoctions developed in a laboratory off farm; the key is creating an on-farm habitat conducive to naturally positive microbes.

Recent discoveries about the human microbiome certainly bear this out. Goodness, who would have thought even 20 years ago that we would be inserting human poop from healthy intestines into a sickly person as a medical treatment? The anti-microbial sterility cult, fortunately, is now balanced with renewed interest in the microbiome specifically and the role of life-sustaining microbes generally.

If we can figure out a way for our housed animals to live on a compost pile, then we have a hygienic habitat for proper immune function. This brings us to the concept of deep bedding. Gene Logsdon wrote about this eloquently in his naughty-titled book *Holy Shit*. The whole book is devoted to the construction and use of what he calls static bedding. I've tried to spice up the concept a bit by calling it a carbonaceous diaper.

Manure and urine are full of nutrients that vaporize if dry and diffuse if wet. Generally, when we talk about housing animals, it's for winter inclement weather. At that time of year, the soil is dormant. The microbes hibernate in the cold, awaiting spring's awakening call. That means nutrients encountered in the winter often either vaporize into the air (we all know what that smells like) or leach into the ground water. Leaching is more insidious, often undocumented because we can't see it or smell it.

This carbonaceous diaper epiphany came to me when I was a teenager. Our gullied rock pile of a farm began responding to our grazing techniques and fertility started increasing. At the time, we fed our handful of cows in the barn on a concrete slab. They ate through a slotted gate that we would move through the hay pile--at that time, we made loose hay with an old-fashioned hay loader. To keep the cows from getting filthy, we would shovel (yes, by hand) out the manure from the concrete slab each week and spread it on the field.

After a couple of years we noticed a pattern. What we spread in January and early February did not make any difference when grass greened up in the spring. But what we spread in late February and early March made the grass grow faster, more lush, and deeper colored. Same material, same cows, same field. The only difference was application time.

Months of cogitating and head scratching later, we realized that the timing difference was in fact the secret. What we spread earlier in the winter either leached or vaporized by spring microbial wake-up. What we spread much closer to microbial wake-up did not have time to vanish and actually got metabolized by the soil microbes. We learned if you try to shove nutrients into the soil when it's dormant, most of them will not be metabolized (eaten).

The soil can only eat when the microbes are awake and lively. Once we realized this, we started thinking about how to store the manure in the barn without making the cows filthy. We needed to stockpile these precious nutrients long enough to get us to spring warm-up. Of course, if Gene Logsdon had written his book back then, maybe we wouldn't have stewed about this as long as we did. Eventually we realized we needed a carbonaceous diaper to hold these volatile nutrients in suspension until the soil woke up.

We bought a used commercial woodchipper. The Arab oil embargo sent Americans fleeing to wood stoves and we were no different. With a big old leaky farmhouse to heat, we needed lots of firewood. That generated lots of branches. When we cut firewood, we stacked the branches with all the butts facing one way. When we chipped, this console of branches was within arm's reach and we could feed the chipper efficiently.

With the wood chips to soak up the urine and provide a diaper, we could go the entire hay feeding time without cleanout. In the spring, when the cows returned to grass, we cleaned out the bedding and spread it on the fields. The soil responded and we thought we'd arrived to ecological nirvana. But then I read Rodale's *The Complete Book of Composting.* That's probably still my number one recommended book for aspiring farmers and homesteaders. Originally published in 1960, this compendium, threadbare, dog-eared, and scribbled in, is still my go-to Bible for how to practically operate a carbon economy. It gives relative carbon:nitrogen (C:N) ratios of a host of farm-generated materials in order to formulate the 25-30:1 ratio that drives successful commercial composting.

Once I digested (pun intended) that book, I took our deep bedding practice to the next level: composting. We still did not have a front end loader and I did all the bedding by hand. I'd shovel off the chips from a flat bed truck. We stored them next to where the cows ate. Every couple of days I'd run the cows out of the barn and with a silage fork and wheelbarrow, add a couple inches of dried wood chips. As this built up, it fermented because the cows tromped out the oxygen. The fermentation kept it from freezing. If I dug into it, it smelled like silage.

The barn smelled like fresh wood chips and the cows loved it. The soft, relatively warm deep bedding offered first class service when blizzards came and the thermometer dropped. Still today, few things fill my heart with contentment as much as going out on a cold, snowy morning and seeing all those cows bedded down contentedly chewing their cuds on that clean, warm bedding pack.

Wonderfully, this deep bedding exudes all sorts of fermented microbes that help maintain udder, hoof, and digestive health. The documentation on the livestock health benefits of deep bedding with a proper C:N ratio leaves no doubt that this is THE way to house animals.

But here's the rub: how do you design a structure to handle the depth and how do you acquire and handle all that carbon? To make sure you haven't missed the obvious, remember that you're going to have at least 12 inches of carbon in your livestock housing unit, whether it's a brooder, goat barn, cow shed, or horse stable. That depth offers enough thickness to create a microbial community.

As this bedding pack decomposes, ferments, or whatever, you will add new carbon material on top. When it gets soiled, you'll add some more. Ideally, you want to be able to go as high as 4 ft. but that's pretty high for some things like a brooder. Minimal accommodation would be 2 feet in height. Some of the height depends on how much manure and urine you're going to accumulate. Remember, you don't want to clean it out until spring. The bedding accumulation obviously is more if you start housing in December versus starting in mid-January.

From a design and construction standpoint, this depth is critical because most farm buildings don't have the ability to accommodate bedding 2 feet deep, let alone 4 ft. deep. If it gets even toward 12 inches, the bedding is up on the sill plates or pushing on the wall paneling. Deep bedding creates a lot of side pressure due to compaction. As the animals press it down, it tends to squeeze to the outside.

In Denmark where the loose housed pig production model was refined, they developed the pony wall idea. A pony wall is simply a non-rotting wall high enough to handle deep bedding. Your sill plate goes on top of that. In Denmark, their hoop house structures anchor atop these pony walls. The pony walls serve a dual function in this case. The first is to hold the bedding in. The second is to keep the pigs from tearing up the hoop house fabric.

We've used different materials for our pony walls, from concrete in the brooder (to keep rats out) to oak boards in the hoop houses. In the pole barn where the cows are, we simply affix pine poles to the barn poles. They rot about every 10 years, but they cost us nothing but a little time and do the job just fine. These crude rustic walls go up 4 ft. and above them we string several wires to keep the cows from falling out of the barn when the pack gets deep.

Our favorite floor material is concrete. If you're scooping material off the ground with a front end loader, nothing beats concrete. But we went many years without it; it's a functional luxury. The other nice thing about a concrete floor is that you can prove that you're not contaminating ground water with your bedding pack. Invariably some bureaucrat or naysayer will come along and accuse you of polluting; with a concrete floor, you can see any leaching or contamination coming out and it's obvious that nothing is leaving the bedding pack.

This whole approach is virtually impossible in iconic and nostalgic post and beam barns. Whether built into the side of the hill--bank barns--or simply built on loose-laid field stones, these barns never have base timbers high enough off the floor to handle deep bedding. That is why I always advise people who have these old post and beam barns to never use them for livestock. Figure out a creative use for them that doesn't involve animals. Farm store, farm school, miniature golf, restaurant, event space--these beautiful old structures can be used for a lot of things. They should never be used for animals. Many should be torn down and repurposed into modern and more functional structures.

As a rule of thumb, if you smell manure, you need more carbon. Now, what kind of carbon? You can make carbon yourself with a chipper but you can also buy it. In our area, the going rate is about $15 per cubic yard. That's bark mulch, sawdust, or peanut hulls. We're not in cotton country, but there you can get cotton ginning trash. In Iowa you might get ground up corn cobs. Corn stover can be used as well. Straw used to be the product of choice, but the horse industry now uses most of that and the short-stemmed grain varieties used now produce almost no straw anyway. The straw market is largely gone in most parts of the country.

For best function, you want your bedding to be as close to 25-30:1 in a C:N ratio as possible. Different carbon sources have different ratios. Sawdust is about 500:1. Wood chips are about 250:1. Wood chips from branches with green leaves still attached can be as low as 75:1. Hay is 35:1.

At the same time, manure is different. Cow manure is about 18:1; chicken 7:1, horse 20:1, hog 15:1. Of course, animal excretions are also different. A cow excretes far more than a chicken per day. You can get as meticulous as you want measuring out your carbon based on excretions per day, but here are some rules of thumb.

First, let your nose be your guide. If it smells like ammonia, that's nitrogen vaporizing; add carbon.

Second, look at the cleanliness of your animals. If they have manure on their sides from lounging in manure, it's time to add more carbon.

Third, because wood products are high in carbon, use some junky hay about every fourth bedding to keep your C:N ratio closer to ideal. When you begin the composting process, you'll be thankful for the soft hay pieces because the pile won't heat up as high or as fast. In the hay feeding area, we simply throw in a bunch of junky hay every fourth bedding; the cows eat about half of it and poop on the other half. The one place where we use no hay or straw is in the brooder, where we religiously use wood shavings. They are curved and friable, enabling even the smallest chicks to dig in them and keep them aerated.

Fourth, store your chips under cover. It can be a plastic cover or a shed, but the idea is to let them go through an initial heat to dry them down. Depending on their C:N ratio, they will develop varying degrees of mold fuzziness on them. That is exactly what you want. It softens the material and creates much more capacity to absorb urine, where most of the nitrogen is. If you store your carbon outside where the rain goes through it, the absorptive porosity of the material is taken up with water and it can't absorb urine. That defeats the whole purpose of the deep bedding and you'll end up using twice or three times as much carbon to accomplish the absorption.

Fifth, when you start your bedding pack, you'll think you can't keep up with the carbon initially. You might even get tired and frustrated for the first couple of weeks. But as it builds and begins to work and mellow, you'll find that it takes less and less carbon to accomplish the same sanitation. When it gets deeper than 24 inches, it'll take half as much carbon to accomplish what was required at 12 inches. That's because the thicker sponge has more microbial activity, which is also using up the moisture. When those cows or goats pee, the thicker sponge drinks it in by the gallon and you'll be amazed how little carbon you need to add to keep things aromatically and aesthetically sensually romantic.

Housing square footage requirements vary with the size of the animal. In general, I like the following amounts (all in square feet):

Cow	35
Pig	15
Sheep	15
Laying hen	3

Before leaving this topic, I have two more issues that are important. The first one has to do with a sacrifice area. Often you don't actually want to lock animals inside in inclement weather or during the whole winter. They often like to go out on sunny days just as much as you and I do; even if it's cold. This outside spot adjacent to your bedding pack is called a sacrifice area. This is an area you intentionally let the animals tear up, like a barnyard, that accommodates their yearning to step outside.

Although we feed our cows in the hay shed on deep bedding like I've described, they have adjacent small sacrifice lots to let them be outside if they want to be. When the weather is bad, they prefer to stay on the deep bedding. On chilly nights, they also prefer the warmer deep bedding. But on a sunny February day, they definitely enjoy going outside for a few hours to soak up sunshine. These sacrifice areas normally get muddy and torn up. That still beats the soil damage and resource loss that would occur if they were out all the time.

As long as the animals are removed from these areas at spring green up, these lots don't suffer permanent damage. In fact, we've grown garden vegetables in these highly fertilized, torn up areas. You can scarcely till up an area better than animals will do it in the winter. If you don't want to plant garden there, succulent weeds will automatically grow to cover it up. The key is to completely remove animals from that lot as soon as grass begins to grow and keep them off for the season's duration.

The principle is still the same. If you overexpose an area, like this sacrifice lot idea, you need a longer rest period to compensate. It's all about balancing out the timing. Long stay, long rest. Short stay, short rest. All relative, of course. And while animals like to go out, even on extremely cold winter days, they don't need a large area. Restricting their access to the rest of your pasture will protect it during that vulnerable winter time, give it time to rest, and incentivize the animals to spend more time lounging on the carbonaceous diaper.

Some mud and crud during a critical part of the winter to accommodate an animal's outside yearnings is fine as long as it will rest a long time between sacrificial use. The sacrifice area is not there to satisfy ecological needs; it's there to satisfy animal happiness. To be sure, some animals are more conducive to a sacrifice area than others. On our farm, we only give sacrifice areas to the cows. We don't offer it to pigs or chickens. One of the reasons is because housing pigs and chickens is far cheaper per pound of animal than cows. The infrastructure per pound of animal is cheaper on the smaller stock; as a result, we can more cheaply give them more room per pound. To offer the equivalent housing room per pound of cow would be prohibitively expensive.

Finally, a word about diversity. In addition to a host-free period, stacking two species in the same area effectively confuses pathogens. In the real world of pathogens, they normally hatch in poop and look for a good place to live and multiply.

Imagine a pathogen coming out of a rabbit and looking around for a partner. Harry the rabbit pathogen spots a hot babe named Matilda and proposes to set up housekeeping on a beautiful hill yonder. It's actually only half an inch away, but in the microbial world, a half an inch is about as far as their poor little eyes can see. Matilda thinks Harry's a hunk, so she agrees, they marry, and they head over to that hill yonder. After a trudge that takes 5 years in microbe time, they arrive and immediately have all sorts of problems: skin rash, cancer, infertility.

What's the problem? They inadvertently arrived on a chicken turd rather than another rabbit turd, and these rabbit pathogens cannot survive on chicken turds. Harry and Matilda meet an excruciating and untimely death, sparing both the rabbits and chickens the disease these nefarious pathogens were going to proliferate. Folks, confused pathogens are a good thing. Diversity provides yet another level of pathogen protection; it's not actually sanitation, but it's similar enough to include in this discussion.

Permaculturalists have preached diversity forever, including the stacking concept. This is the idea behind the Raken house, where we house rabbits above chickens. It's the idea of the table-top mezzanine in the hoop house to offer chickens a second level safe haven above small pigs, or sheep, for example. We'll get into this more later, but I wanted to broach the practical implications of species diversity in a housing situation. The conventional view is that proximate multi-speciation increases disease and filth.

On the contrary, multi-speciation deters virulent pathogens by keeping them in check with dead-end material. The deep bedding provides the structural base, but on top of it you can put as many different species as you'd like. We've run rabbits, chickens, and pigs altogether in the hoop houses and it works great. This set-up also enables us to use the cubic footage of the infrastructure rather than just the floor square footage. That's a real game changer if you're trying to get more throughput from your infrastructure.

Here is the bottom line: if you're going to house animals for any reason, they need a carbonaceous diaper. That requires depth, which requires walls that won't rot and that are strong enough to contain the material. If you smell ammonia, you're losing fertility, which we'll discuss in the next chapter.

A skiff of shavings; a dirt barnyard; a short-grazed horse paddock; a pasture that houses a cow for months on end without breaks: all of these court sickness and disease in your animals and destroy the ecology in and around your homestead. If I knew of other alternatives, I'd certainly try them. As far as I know, rest and sunshine or vibrant decomposition are nature's way of keeping things clean and healthy. That means on our homesteads, even at one-animal scale, utilizing nature's template will go a long way toward keeping the vet away and at the same time make your homestead a nest of animal happiness.

Chapter 9

Fertility

Animals are worthwhile on homesteads if for no other reason than enhanced soil fertility. Especially herbivores. That's because they are inefficient. When anti-animal folks scream "INEFFICIENT!" in the face of cows, they're exactly right. The inefficiency is why they move soil fertility forward. If cows were as efficient as cabbage, we wouldn't have any soil on the planet. That is, unless you plowed down four out of five crops of cabbage.

Left-overs build soil. Whether it's direct biomass or decomposed biomass like manure (or compost), food for soil microbes comes from salvage material. Routinely removing all the vegetation from an area soon impoverishes the soil. Nothing leaves as much behind as cows and grass, or perhaps more precisely, perennials.

This is why Wes Jackson at the The Land Institute has devoted his life to trying to find a perennial grain. His research is ongoing but it's painstaking work. Compared to the research that has gone into eliminating the leaves and stems of wheat, barley, oats, rye, soybeans, and corn--the big grain crops--Jackson's lifetime of work doesn't seem like much. Hopefully it will continue far into the future.

Perennials pump energy into the soil. Annuals extract it. This is the basic concept behind the historic seven-year crop rotations that always included four years of perennials. Variations on this theme exist throughout the world, but the general multi-year perennial versus few-year annual is axiomatic to preserve soil fertility.

The subject of soil fertility fills libraries, of course, and I will not bore you with a repeat of all that information. I've read a lot of it and I'll give you the nutshell of what I think. The engine that drives fertility is carbon.

Perhaps no one exemplifies this principle better than Paul Gautschi who wrote the book *Back to Eden* about his deep wood chip composting methods. Just like any visionary practitioner, he has his naysayers. But I've seen many people adhere to his principles with great success. In fact, he's a permutation of our family's earliest successful gardening technique gleaned from Ruth Stout's *No Work Garden Book*. She used leaves and seedless hay as her mulch.

In full confession, we did the Ruth Stout system for about a decade some 40 years ago until we inadvertently brought in some wire grass roots in leaves from the city dump. That stuff would grow up through anything. To prove it one time, I put a clump of it on the ground and then piled leaves four feet deep over top. Sure enough, within a couple of months, that wire grass grew all the way up through that leaf pile.

The only thing I know that will kill wire grass is tillage. In fairness, tillage does not require a tiller. It can be done with a broadfork or spade to loosen the soil and then grub it out by hand. I've done plenty of that. Tillage can even be done with pigs. Certainly one of the central themes of today's regenerative agriculture movement is no-chemical no-till. You have to put the no-chemical in front to separate it from the chemical no-till which has dominated mainline no-till agriculture for 40 years. The point here is that I'm not a no-till cult leader. That said, tillage should be extremely rare.

Nature builds soil without tillage. The deepest and best soils are not under trees, but under prairies (grasses). Grass is far more efficient at converting solar energy into biomass than trees. Trees, however, serve many useful purposes, including being able to hold biomass a long time. Grasses tend to choke themselves out or burn if they accumulate; trees can store carbon longer in a useful state without creating a fire hazard or choking themselves out. For optimal function, though, trees need to be thinned. I like the permaculture axiom that we need more forests and fewer trees.

Nature builds soil without tillage, with carbon, on top of the ground. Grass falls on top of the ground. Leaves fall on top of the ground. As my friend Bill Wolf, founder of Necessary Trading Company in the earliest days of the organic movement, used to say, "earthworms like to be fed on top of their head." In other words, you don't have to till something into the soil in order for it to be useful to the soil. On top is just fine; the worms and other critters like dung beetles will bring it downstairs where micro-organisms can digest it.

Many new homesteaders purchase a run-down place. "The soil is as hard as a rock," is the common first assessment. First, don't think that's different than 95 percent of all agricultural land in the U.S. By and large it's been overgrazed, over chemicalized, over-tilled, pounded and abused for more than a century. You'd get hardened too if somebody pummeled you every day.

What you have to understand is that the soil is biological, chemical, and structural. But these parts don't work independently. They work symbiotically. It's not addition; it's multiplication. In other words, 0 X 50 = 0. If it's hard and you till it to fluff it up, but it lacks basic biology, you won't get anywhere.

If you take a soil sample and it's lacking in potassium, applying that to hard soil won't do you any good. Back in the 1960s a friend of my dad's purchased a train car load of soft rock phosphate and talked Dad into sharing some of it to get the whole load at a discounted price. Soil samples indicated our fields were desperately low in phosphorous. We applied two whole truckloads and never saw one iota of response. Nothing. Our soils were hard as a rock and lacked biology.

One friend told us we needed to till and plant a green manure. We didn't own a plow but we had a horse of a garden tiller. We needed it because the garden was like brick. We didn't know as much then as we do now. So Dad took our big garden tiller up in the field and tilled up a quarter acre and planted a cover crop. It didn't germinate very well and instead of being a fertile spot, that quarter acre was an eyesore for a couple of years.

When we came to our farm in 1961, it was gullies and rocks. I can remember well walking the pastures and being able to step between the vegetation. I could walk the entire farm on bare dirt. Large bare shale spots dotted the fields, some as big as a quarter acre. These were bare rock, like big rock saucers in the fields. At that time, the whole 100 acres could scarcely support 10 cows. To say that it was an impoverished piece of land would be the understatement of the century.

But out of extreme crisis comes extreme innovation. Fortunately, our family didn't have money. The combination of desperation and destitution incubated creativity. Dad discovered Andre Voisin, the founder of modern controlled grazing and author of *Grass Productivity.* He was already familiar with Rodale and organics, but the controlled grazing was new information.

By combining controlled grazing with composting and then eventually with poultry, the soil ever so slowly began to respond. As I grew older and my observations became more acute, I became interested in weeds. We had lots of them. I devoured weed books not only for identification but also for what they indicated about the soil. Most weeds are part of the healing process.

The most dramatic fertility story I have centers around broom sedge and red clover. We'd been on the farm a decade; I was a teenager. It was hard scrabble. We had broom sedge in the fields thick enough that a casual observer would think we were growing it for a crop. In the fall, our pastures turned yellow with broom sedge. I looked up brooms sedge in my weed books and the narrative went like this: indicates calcium deficiency and poverty soil; remedy by applying lime.

We never had any clover, but I knew clover was desirable, so I looked up clover. The narrative read like this: indicates high calcium and rich, healthy soil. All the farmers in our area routinely apply lime to keep calcium up in their fields. Did I mention that our fields were full of brambles? As kids, we would pick gallons of dewberries (ankle scratchers).

One day walking through the pasture, I noticed a strange thing. A red clover blossom in the middle of a broom sedge clump. Half a century later (could I be that old?) I still remember it like yesterday. I was coming into the farm, as they say, and was vibrantly interested in everything about it. I was old enough to start formulating some ideas. This was my burning bush experience.

I kneeled down and examined this strange sight. The red clover only had one frond and on the end a small but beautiful red blossom. It was growing out of the middle of a seemingly robust clump of broom sedge. How could this be? The two plants, according to the books at least, indicated completely different soil contexts. One needed low calcium and the other needed high. One indicated poverty, the other indicated health. And here they both were, occupying the same square inch of ground. How could two species requiring opposite environments live in the same space?

Then it hit me. This is healing, nature's style. It's not flashy. It's not fast. It's gentle, quiet, and beautiful. The drama is in the changeover. Could we have accomplished the change faster by applying lime? Perhaps. But by the same token, the lime may not have moved the needle at all because we didn't have enough biology to use it. Allan Nation used to say that biology has its own clock, and it doesn't run very fast.

In mechanics, we can change parts, cut, weld, fabricate. But nature is subtle and changes over time. I'll never forget that moment when I saw, in real time, the invasion of health pushing out the remnants of sickness. It affirmed that the things we were doing with the grazing, composting, and multi-speciation were working. Stay the course. Hang in there. Nature responds to both abuse and respect. Unlike machines, it can heal from the inside out. A shattered bearing will never heal, no matter how much love and rest you give it. But living things can heal.

Over the next couple of years, the broom sedge literally left. It simply died out, replaced by clovers and productive grasses. We never planted a seed, never spread lime, never plowed, overseeded, or any of the normal recipes. We didn't even soil test. And those big rock areas? Just like a wound, they began healing from the outside in. Each year, the sod would encroach another 18 inches along the outer edges.

In 2021 we will have been on this land for 60 years. At about the 45-year point, the last square foot of those rock areas disappeared under the healing sod. In a severe drought, those areas are still visible because they dry out earlier. But the fact that we have a foot of soil over what were once quarter acre bare rock patches is remarkable indeed. The soil is like skin and it has a capacity to heal and rejuvenate.

What do all these waltzes down memory lane for me mean for you? My intent is to encourage you to take care of carbon and management first. I understand being intimidated by ignorance. Who isn't? The typical sequence goes something like this: homesteader buys property, looks at poor pasture, begins asking advice of neighbors and government experts, pulls soil samples, buys expensive amendments, becomes frustrated with the cost of the homestead, sells homestead. It happens all the time.

The point of sharing my own fertility experience is to change this sequence. I'm not opposed to soil samples or amendments. But if they precede excellent management and an intentional carbon application program, they represent the proverbial cart before the horse. Carbon of course includes composting and is related to excellent grazing management due to the plant's self-pruning of root mass to mirror above-ground biomass. This pulsing of the pasture is like a carbon heart beat.

Once you get the carbon and animal control developed, then you're ready to pull soil samples and tweak with amendments. You may see enough improvement that these refinements aren't necessary. You may not.

Herein lies an important lesson: it's all about your starting point. Let's use the metaphor of a race. You have three competitors lined up at the starting block. One has a broken leg; one never exercises and is morbidly obese; the third is in good shape but lacks some training. If the goal is to run the race in one minute, each of these will have a different regimen for success.

The fellow with the broken leg needs a splint, perhaps some physical therapy, and time to mend before he can even start running. The obese guy needs to lose some weight and really start moving around. The third guy simply needs a good training regimen and perhaps a coach. A running coach for the first two is inappropriate. The point is that each of these competitors requires a different program. All the programs require training and exercise, but what can be done day one, day two, and so on varies depending on the starting point.

Now imagine each of these potential runners is a fertility situation--or lack of it. We could call the race a fertile soil. Depending on the level of handicap or dysfunction, the starting point varies and that makes the regimen different. Climate can play a big role in this. Generally, soils degenerate slower in colder, drier areas. In warmer, wetter areas soils can degenerate faster.

The reason is that the shorter growing season also shortens the ability of a farmer or rancher to harm the soil. Because decomposition is slower, the virgin organic matter inventory holds up longer. In warm and wet areas decomposition occurs longer and more rapidly. That's why, for example, ecosystems in the north (at least in the northern hemisphere) hold up to 80 percent of their organic matter in the soil. In the tropics, this ratio is inverted and 80 percent is above ground.

That's why when farmers clear tropical forests for cropping, the soil loses its tilth and turns to brick in a couple of years. In cold areas, the soil is more forgiving because the ecosystem stores most of its organic matter below ground. A further issue is the period of time an area has been abused. For example, our farm was settled by the Europeans in the late 1700s. Although the Native Americans practiced the three sisters cropping system--squash, corn, beans--they did not have steel to rip through the root web of prairie grasses.

The natives' primitive tools and the grass root web made tillage extremely difficult, putting a labor boundary around what could actually be disturbed. Their most disturbing tool was fire and they used it routinely. In our area, the Europeans largely enjoyed a firescape landscape developed and nurtured by the Native Americans; a magnificent silvo-pasture that literally built the deep, alpha soils that dominate the Shenandoah Valley. By the time our family arrived in 1961, roughly two centuries of European tillage damage sent 3-8 feet of topsoil off the hillsides, down the rivers, and out to the Chesapeake Bay. Hence, the exposed rocks when our family arrived.

If we go back to our running metaphor, our farm was not the fellow who simply needed some training; it was not the out-of-shape obese fellow; it was the fellow with the broken leg. In fact, our race started with two broken legs and an auto-immune deficiency, leaky gut and irritable bowel syndrome. I mean, this place was in rough shape. Ohio was settled by Europeans a century later. Those soils have a century less of abuse than the ones here in Virginia. As you move north into the Dakotas and Canada, not only have the soils been abused a shorter period of time, but the shorter growing season means they can't be assaulted as many days per year as those in the deep south.

In general, the farther north you go, the faster your soils will respond to a good regimen. What took us 40 years to accomplish might only take 5 years in North Dakota. Same regimen; different starting point. So on our farm, we imported corn cobs, leaves from the city dump, compost from a local live animal auction yard, and even took in walnuts for a Missouri nut processor. Walnut hulls are an excellent source of carbon. So are coffee grounds if you happen to live near a roaster.

Some folks ask us why we don't do keyline plowing. I'm a huge fan of keyline plowing, developed by P.A. Yeomans the Australian water wizard. But if we ran a subsoiler or keyline plow on our place, we'd have nothing but rocks--big rocks-on top of our fields. We've simply stayed with the carbon on top of the soil, excellent animal management, and time. The earthworms and soil micro-organisms took that management and carbon and worked their miracle.

When I began driving the tractor to rake hay as a young teen, I remember scraping together 16 swaths to make a paltry windrow. Today one swath makes a windrow I can't jump over--and it's not because I'm old and fat. Ha! Fields where we used to make 80 bales now make 1,600. This is not bragging; I'm trying to inspire you to think about fertility in an overall context of resource management.

Over the years we've played around with everything from paramagnetic rock to biodynamics to foliars and nothing makes a difference as fast or comprehensively as management and carbon (compost). I know some folks cringe at my interchanging carbon with compost because compost is certainly far more than carbon. But I put them in the same general category as a soil fertility builder. Goodness, just spreading leaves a couple of inches deep on a poor pasture will yield fantastic results in one season. Carbon feeds soil. Compost is better, but carbon is good enough.

If you buy sophisticated soil amendments and spend your time pulling soil samples and poring over lab tests when you should be building fence and negotiating with sawmills and vineyards (pumice) for carbon, you'll spend a pile of money and not get where you want to be. Fertilizing a mismanaged pasture is like taking all the money in your wallet and throwing it in the fireplace.

To be sure, one of the most common ways to import fertility is with feedstocks. Whether it's buying in hay for herbivores or grain for omnivores, this injects significant carbon and its cousin, manure, into your system. On a small acreage you can never justify owning hay making equipment; probably not even a tractor. Buying in hay can offer a splint to the broken leg (back to our running metaphor) and push the healing forward.

In my experience, nothing moves fertility forward as fast as chicken manure. Pastured poultry, and especially pastured broilers with their intensity, can create fertility out of poverty in a couple of months. People routinely ask me if we had it to do over again, could we have moved our fertility forward faster? Absolutely. I have no doubt that we could have done what we did in 40 years in only 10 had we known more at the start. That's what this book is all about--compressing your learning curve and hastening your success by tapping into our family's trials and errors. For a more comprehensive look at this whole story, please read my book *The Sheer Ecstasy of Being a Lunatic Farmer.* I call it my soul book.

Utilizing your animals as a conduit for fertility development is perhaps the greatest leverage you can derive from livestock. Rather than something that makes stinky barnyards, packed down loafing areas, and overgrazed fields, animals are the greatest catalyst for soil development if they are part of an overall carbon and control strategy. Absent that, they are the most effective destroyers of soil and ecosystems.

What about bringing in compost or manure-type materials, like poultry litter from factory chicken houses? This is a natty discussion, especially with purists. The fact is that many of our homesteads are in triage. You do things in triage that you wouldn't necessarily do when everything is healthy and functioning. I see a big difference between priming a pump and ongoing dependency.

The whole goal of priming a pump is to keep from having to do it again. The whole point of putting a splint on a broken leg is to eventually dispense with the splint. But if the pump is dry and the leg is broken, you do non-ordinary things to get back to smooth, independent function. With that in mind, I have no problem bringing in material of dubious provenance to prime the pump on your fertility program. Don't let perfect be the enemy of good enough. Sometimes you can't get perfect, or you can't afford perfect. Good enough is perfect in this case.

If perfection bankrupts you, what good did it do? As a pump priming scheme, go ahead and use the cheap, locally available material. It might be confinement house cleanings; ideally, composted. The problem is that even when these factory farms say "composted" it's not. Oh, it goes through a heat period, and that's helpful, but it's not really compost because they never put enough carbon with it to balance out the nitrogen. It still burns your nose when you inhale up close. That burning is the vaporizing ammonia that contains the nitrogen. You can stick your nose up against true compost and inhale without a burning sensation.

Ideally, you'd bring in this material, mix about 25-50 percent carbon with it (by volume), add water, and let it go through another heating (composting) phase. But again, ideal may not be possible or may be more trouble than it's worth. Go ahead and spread it. But if your fertility program depends on the injection of this dubious material each year, you have a problem. It's not balanced and will eventually create soil problems. Furthermore, you're importing antibiotics and other things you might not want proliferating on your homestead.

If bringing in factory farmed wastes primes your pump, go for it. But only once. And if you bring that in before having your control mechanisms and carbon plan in place, shame on you. Imports, amendments, soil sampling--it all follows control and carbon.

If you read about a magic fixer elixir and want to try it, do it on a small scale. Some of the amendments I mentioned above we've tried on a small scale. Realize that what works in one area may not work in another. Remember our metaphor about the runners. If the training coach of the third runner went over to the first one with the broken leg and said "You need to run down to that tree and back as a starting point in our training," it wouldn't help; it would hurt. All progress takes an incremental progression.

Don't just assume that what worked one place will work someplace else. The only thing I know that works everywhere on the planet is control and carbon. Other things may, but they may be more costly than they're worth. When I say something works, I mean it works not just ecologically, but also economically and even emotionally. Emotionally includes hassle and noxious factors. I'm a big fan of worm exudates, but somebody has to set up the infrastructure and feed the worms and extract the substances to make it work. Ditto compost teas and other home made foliar amendments.

I love them all and encourage experimentation with all of them to add to our collective body of knowledge. But always play around with it on a small scale. Do side-by-side trials. As a rule, about 80 percent of the decisions farmers make are based on the advice of a salesman. Run everything through your experimentation program. But start experimenting only after you've implemented your carbon and control system.

If you're disappointed that I'm not giving you a comprehensive green light, red light on all the fertility programs and amendments out there, it's not because I don't have opinions. It's because I've done enough experimenting and visited with enough folks in their farm house living rooms to know that it's a big world with a lot of options, a lot of gurus and way too much genius for me to sort out. And we have new materials and protocols every day that make current recommendations outdated by the time I could get this book into your hands. No, the answer is principle and overall philosophy. If you have that right, the nuts and bolts will work themselves out on your place.

One of the most dramatic fertility responses we ever saw was one year when we accidentally purchased an incorrectly ground batch of kelp. We use kelp meal religiously as a mineral supplement for our cows. When we brought the bags home, instead of the coarse ground material we normally got, it was powdered. What we got was the grind used for soil amendments. It was not our fault; it was incorrectly labeled. I didn't want to return them to the store because I hate to complain. I think that comes naturally to folks who are in business for themselves.

We tried to use it for the cow mineral, but they wouldn't touch it since it was like talcum powder. They couldn't swallow it. I finally decided to add a bit of it to each spreader load of compost I took to the field. This was in the early days before our fertility really started to come up. Those six spreader strips through the field were noticeably greener and thicker compared to the adjacent ones that didn't have the kelp (after I used up the six bags). I looked at those strips and dreamed about buying enough for all the compost, but I didn't have the money.

As a serendipitous experiment, however, it was fascinating to see what adding the kelp did to the compost. Today, we have an outdoor wood furnace and add quite a bit of wood ashes to the compost when we spread it. We also add Planters II trace minerals when we have enough money. If we have some extra cash, we add the mineral; if we don't we just use the ashes because they're free. Wood ashes might be one of the most overlooked sources of minerals.

Interestingly, a friend purchased some property close to us a couple of years ago and wanted to get a base line soil sample before he did anything. He asked if he could take some samples on our place to have some comparisons. The lab is known for its mineral advice; I've never seen a soil sample come back from there without recommending mineral. The samples from our fields came back: nothing recommended.

I couldn't believe it. Without any amendments except control and compost over the years, we have saturation even of calcium. Without ever applying lime. Oak bark contains large amounts of calcium and our wood chips are largely oak. No doubt we've been adding a lot of calcium with our chips.

And this brings me to yet another point: carbonic acid. That is the acid produced when carbon dioxide combines with water in the soil. If I put a rock on the table and want to know what minerals are in it, I can treat it with numerous acids: sulphuric acid, hydrochloric acid, etc. Guess which one works the best? You got it: carbonic acid.

Many times our soils have enough parent minerals in them, but they are inaccessible because we don't have enough decomposition (there's that carbon again) going on to make enough carbonic acid to break out the minerals. This is just one limitation of soil samples. Many experts recommend plant tissue samples rather than soil samples. In my opinion, none of that will ultimately be your weak link. Your weak link, and mine, most of the time is lack of control and carbon. Fix that, and fertility begins to improve. Soil samples taken from the same spot over the course of a growing season indicate substantially different saturations.

I still remember attending a session about soil fertility at a conference many years ago. The Virginia Tech PhD agronomist wrote three things on the blackboard (this was before powerpoints): C, O, and H2O. He introduced the session by saying that these are the three most common substances in soil. "But we're not going to talk about them; we're going to talk about N (nitrogen), P (phosphorous), and K (potassium).

I remember sitting there thinking that if he would talk about carbon, oxygen, and water, he probably wouldn't need to talk about N, P, and K. But such is the world of soil science, or at least it was at that time, that the real issues took a back seat to things the fertilizer sponsors lining the vendor booths at the back of the conference hall were selling. Am I the only one that smells collusion here?

One final point on fertility: irrigation. From our previous discussions, you know I only advocate irrigation from impounded surface runoff. The New Zealand K-line system is my favorite because it can be purchased incrementally and it's conducive to any configuration you can imagine. We have several lines that together can cover a substantial amount of our pastures. We've run electricity out to several ponds to power electric pumps that operate quite cheaply.

Because pond water is soft, warm, and full of microbial life, it's the closest to resembling rain; far more similar than aquifer water. In a drought, keeping biomass growing is a critical part of carbon. Capturing surface water runoff is part of your overall resource control strategy. See how it all works together? Controlling surface runoff works hand in glove with controlling the manure resource and managing the animals as pruners. This is all part of control which in turn is critical to synergize the carbon cycle.

Be patient. Be methodical. Above all, don't twist yourself into a pretzel based on the advice of neighbors and experts. Trust nature. Trust experience. The weak link in fertility is almost never something you can buy. It's something you have to do and develop on your own acreage. Control and carbon.

Chapter 10

Developing Pastures

"What should I plant for my pastures?" is perhaps the most common question from new homesteaders, other than "What breed should I buy?" At the risk of irritating every seed company and purebred operator out there, the answer is "Nothing . . . at least for now."

It all depends on your starting point, which is similar to the discussion we had surrounding fertility. The most common starting point is a worn out pasture. The vegetation may be sparse and spindly, lots of weeds and poor fertility. I think we've established that investing in seeds prior to dealing with fertility is the wrong order.

Succession is one of the most powerful forces in nature. Nothing living stays the same day to day or even minute to minute. All biological entities go through birth and death. Beyond that, though, the plants and animals inhabiting a given square foot or region manifest the management record.

Today's vegetation records what went on before. A field of corn, for example, reveals that at some point someone plowed the sod and planted corn seeds. Looking at the soil may tell us how long ago that occurred. Done over a big enough area for a long enough time, the corn mono-crop and that initial tillage extends its reach into the spider population, moles, voles, red-winged blackbirds, indigo buntings, foxes, rabbits, earthworms, fungi, and even cloud formation.

Landscape activity has a domino effect and etches its diary into the ecosystem in ways we're still discovering. Learning to read the historical legacy from vegetation is of course a life-long skill. The longer-lived the vegetation, the farther back you can reach. For example, our farm has a piece of hardwoods littered with dead pines. We've heard based on local lore that this woods section was a grain field during the mid 1800s. The forestal record bears that out. In our area, pines need full sunlight to thrive; that's why they're called pioneer species. After they initially overtake an abandoned field, hardwoods sprout in the understory.

Hardwoods can thrive on 50 percent sunlight; limiting the sunlight helps these hardwoods grow tall and straight. In the natural succession, the pioneer pines act as a nurse crop to the fledgling hardwoods. The pines, and especially Virginia pine, are short lived, succumbing to bugs by 50 years old. By this time, the hardwoods are 20-30 feet tall. As the pines die off and fall over, the hardwood canopy fills in and the climax forest expresses itself. Eventually, of course, even those trees die, opening up new opportunities for young ones.

The point I'm trying to get across is that as you walk across your field, it's not a static thing. Even if it looks poor and stagnant, it will respond to resuscitation. The most important thing to remember is that the soil seed bank is enormous. Nature's seed inventory is measureless and abundant. Not only do seeds arrive on your doorstep every day, carried by wind, feathers, hides, spider webs and fur, but they last a long time in the soil, buried for future awakening. When you plant a garden, do you plant weeds? Of course not. And yet we have no shortage of weeds; obviously those seeds are there in the soil, waiting for the right circumstances to sprout.

Christine Jones, the Australian soil guru, has a wonderfully whimsical way of explaining this seed inventory. She talks about the wisdom of the seed to know when to sprout. Some clovers, for example, stay viable for 1,000 years. Imagine a clover seed falling to the ground this year. Next year it could sprout, but it decides not to. Ditto the following year. Ditto the following year. I won't draw this out, I promise.

At year 100, it continues to lie dormant in the soil. Year 200 comes and goes. Year 300 passes. Something happens to its surroundings in year 997 and it responds "this is the year I'll sprout." Perhaps that's the year you applied some compost and ran six cows across the square inch where that seed lay. With a new context and new prospects, the seed decided to germinate after all those years. Perhaps that is what happened to the clover seed I described in the previous chapter that pushed its way up through the broom sedge.

Under your worn out pasture lie millions of latent seeds waiting for things to change. They're awaiting the "all clear" signal, or the "it's okay to come out now" invitation. That a seed carries that much wisdom should give us all pause about swashbuckling across the landscape with a pirate persona. This is why we caress the landscape rather than wrestling with it.

Whenever anyone asks me what to plant, my first response is always "what's in the side ditches." While you may find plants that have performed better in production trials, studies comparing all energy in to all energy out point to the natives as the best overall performer. What good is a top production forage if it's finicky and requires unnatural amounts of nitrogen to be healthy? Some land grant colleges developed forage varieties with one purpose in mind: metabolize as much factory farmed animal waste as possible.

No natural system generates that kind of excess nutrients. The problem with these specialty sophisticated varieties is that they don't do squat unless they get copious amounts of inputs.

Researchers many years ago measured total inputs versus total outputs and found that the road side ditches offered the best option. New Zealand forage specialists add animal performance and palatability to the production-only criteria and find that often the most productive plants are not the most palatable.

What good does it to do produce a mass of forage that the cows don't like? These resilient native or nativized (my word--it means they've been here a long, long time) forages illustrate their hardiness by thriving with little to no care. Just a little attention brings them to their full potential.

How do we get this native and latent seed inventory to germinate? Disturbance. While fertility certainly has much to do with what vegetation grows on a given spot, disturbance is the ultimate catalyst. I call it ecological exercise. Vaughan Jones, long time New Zealand farmer and author, called it "deep massage."

Let's not get hung up on terminology; what we want to understand is principle. Disturbance comes in many forms: mowing, fire, extreme weather (cold, hot, wet, dry), tillage, and hoof action. The one we want to zero in on here is hoof action. If you've ever had a cow or horse step on your foot, you know being trod on is disturbing. In arid areas, it uncaps the soil to let air and water penetrate. In more temperate areas, it can cause pugging in extremely wet conditions.

In general, unfrozen wet ground in the winter is far more fragile than in the summer, primarily because nothing is growing. On our farm we've pugged up areas bad enough that from a distance they looked freshly plowed. The first season after pugging the area grows nothing but weeds. But the second year we've had these areas grow pure timothy, a highly valued, palatable, and beautiful grass. We never planted a seed. The timothy grew from the seed inventory and apparently all the conditions necessary to awaken those types of seeds converged at that time.

The close cousin to disturbance is rest. Those familiar with Allan Savory's ground breaking work will recognize these two principles as foundational for ecological succession. Disturbance normally needs to be quick and the period of rest varies in length. You never want long disturbance; the longer the disturbance, the longer the rest must be.

The interesting thing about disturbance is that you can't always see it readily. For example, a five acre field with two cows in it year-round might not look disturbed. But individual plants that the cows like I can assure you are way over-disturbed. We could call this over-grazing. They are disturbed over and over and over again. We'll get into this more in the upcoming grazing discussion. For now we need to appreciate that disturbance is not always apparent. A cow path is obviously over-disturbed to the point of erosion, compaction, and zero productivity.

The reason I promote animal movement from paddock to paddock is not just for sanitation; it's for rest periods between disturbances. Again, if we think of disturbance as exercise and rest as the relaxation between exercises, it helps us to appreciate what's going on ecologically. Rest is a time of recuperation. Disturbance is a strain on the system, but that's what builds muscle.

The reason a pasture is bedraggled and worn out can be a result of too little exercise and rest repetitions. Leaving animals on a given area pounds lounge areas, walking areas, and palatable plants daily without rest. Over time, we get high impact zones and dead or weakened plants. Just like disturbance takes many forms, rest does too.

One of the best things you can do to jumpstart a lethargic pasture is to let it rest long enough for most of the plants to make seeds. Weeds too. In our vernacular, we call this letting it blow out. The plants get stemmy and turn brown as the seeds shatter and spiders build webs between the tall spires. Putting animals on such a spot, at extremely high density, for one day, stomps that lignified carbon onto the soil surface, pushes the seeds into the ground, and lays on some much-needed manure.

Jim Gerrish, grazing management expert, preaches the value of wasted grass. Whether it's a gentle graze on pre seed head grass or a late graze on blown out seed head grass that tromps half of the biomass onto the ground, uneaten grass is what ultimately feeds the soil. Occasional pounding is highly beneficial because it opens up the sward to awaken new sprouts of different plants.

Having established the value of disturbance and rest in vegetative succession, the problem facing the small holder is not enough animals to create the power of an excited mob. Said another way, 5 cows on a twentieth of an acre are not the same as 100 cows on an acre. Same density, but extremely different impact. The larger group inherently moves more like a crowd rather than a group of individuals.

One of the most interesting side-by-side comparisons I ever did was one winter when we had a lot of extra stockpiled grass going into winter and not enough cows to eat it all. We only had 20 cows at the time. We pushed a pencil on different scenarios to eat the grass. We could buy older cows and sell them in the spring. We could buy stocker calves and sell them in the spring. We knew the extra grass was an anomaly: we'd had a rare extremely warm and wet fall, creating an explosion of growth. The extra fall grass certainly did not guarantee extra grass in the spring.

We knew this was a short-term problem and eventually teamed up with a neighbor who had 400 cows. He brought 80 cows over to our place on a per-diem rate; with this scheme, we sold our standing grass as a cash crop but harvested it with cows and enjoyed the benefits of all their manure. We didn't want to run those cows with our cows for a number of reasons.

We had a 12-acre field and started our small herd at one end in a quarter-acre paddock and the neighbor's larger herd in the middle in a one-acre paddock. Both groups headed west in the field to keep them well apart. The big group needed 30 percent less ground per cow than the small group. It was my first time running a group larger than 50 and the differences were profound. The larger herd moved and acted like one big organism; my 20-cow herd, even when compact, acted like 20 individuals.

Since then, we've run groups as large as 720; that's a rush. As groups approach 100 head, they became like a giant amoeba, moving like a mob rather than individuals. I wish I could tell a small-holder how to get mob action out of a handful of animals, but I don't think anyone has been able to do that yet. The point is that the disturbance benefit issue is easier to obtain with a larger group than a smaller one.

Still, you can do things to simulate a larger group disturbance. One is with the mineral box. This is why you want a highly portable box. If a mineral feeding system is hard to move, you won't locate it to greatest strategic advantage. I'm going to belabor this mineral box a bit because it is one of the most underutilized techniques on livestock farms.

Think about the natural flow of nutrients. Even if your homestead is fairly worn out, I guarantee that you have spots where fertility is high. These might be in a valley where minerals, manure, and pieces of vegetation naturally collect. It might be a ridge top where animals tend to lounge. In any case, no matter where you are, your place will have large variations in fertility and vegetation quality.

Always place the mineral box in a spot that can use some manure and trampling. It might be a spot where the grass is thin; it might be a spot of noxious weeds; perhaps a spot of brambles. Be highly selective about box placement. It attracts the animals and they often jockey for position around it, creating a bit of a ruckus that translates into haphazard hoof action on the soil. That's disturbance.

Never place the mineral box where it's convenient for you. If it's convenient for you, it's probably convenient for the animals and if it's convenient for the animals, they've probably historically spent plenty of time there. Always place it on the rocks, the slope, the weeds; never a ridge top or a valley that more naturally enjoy animal presence. One of my pet peeves is seeing a mineral box located in a valley, but more times than not, that's exactly where farmers place them. Why? Because it's easy to access. But a valley always has the best vegetation in it anyway.

Place the mineral box away from access areas and gates. Again, I often see the mineral box placed right inside a gate. Folks, the animals naturally disturb access areas; if any place is torn up on your homestead, it'll be the areas right around gates. Never place the mineral box there. Get it out, away, where the animals don't naturally spend a lot of time. You're trying to draw their manure and their hooves onto unfrequented places.

The next high impact disturbance technique is water trough placement. Similar to the mineral box, this is why we never want stationary water placements. If you visit an abandoned farm with stationary water placements, you'll notice the rich dark green color and density of the forages near the drinking stations. Over time, the disturbance and nutrient load combine to create highly fertile areas there but they can only be expressed after extended rest. As long as the water point is being used, the animals over disturb and over fertilize the area, turning it into a mud hole when it's wet and a dust bowl when it's dry.

For the same reasons we want portable mineral boxes, we want only portable water troughs. To be sure, since your water trough needs a hose connector to your main water line, you can't be as free with its placement as the mineral box. But within the limitations of a 50 foot or 100 foot hose, you still have a great deal of flexibility for placement. Even a couple of animals will tear up the area where they drink; moving the trough around moves your disturbance around.

The third major disturbance technique is portable shelter. Concentrating the animals in any way enhances the disturbance factor. We'll talk more about shelter later, but the point here is to see it as a part of the overall disturbance strategy which is the foundation of moving vegetative succession forward. This is why I'm a fan of portable shade rather than trees.

Here is where I'll deviate a bit from my silvo-pasture and perhaps permaculture friends, but hear me out. Just like stationary mineral boxes and stationary water points, stationary shade over-concentrates manure and disturbance in one spot while denying those benefits to the rest of the land base. Whether you have one acre or 5,000, the principle is the same: equal distribution of all the positive successional techniques. This is healing management that the human factor can bring to a landscape.

Nature does this with predators, fire, and massive herds. Absent that, we need to do it with management. While I'm not opposed to widely scattered trees in a pasture, in general I think we can move vegetation forward faster with portable shade. It's also more hygienic because we can change the lounge area daily. Since animals aren't potty trained, they defecate liberally where they shade up. Making sure that each time they come through an area they place their droppings and their stamping hooves in a different spot is critical to the entire vegetation ecology.

The next disturbance technique is poultry. Nothing moves vegetation forward as rapidly as chicken manure and scratching. When graziers (the grazer is the animal; the grazier is the manager) talk about disturbance, seldom do they think about the omnivores, but that is a big mistake. Multi-species schemes simulate the massive flocks of birds known to co-habit the prairie with the different herbivore species. Leaving out the birds (chickens) misses one of the key ecosystem drivers.

Native American lore includes stories describing passenger pigeon flocks roosting heavily enough in trees to break their branches. Naturalist John Audubon recorded in his diary that in one of his wilderness wanderings he couldn't see the sun for three days because it was blocked by a flock of passenger pigeons. Folks, adding chickens to our homesteads is about as basic an ecological foundation block as we can find.

Truly free range unrestricted chickens from our Eggmobile happily go out 200 yards from their home base and return. They will cover a circle 400 yards in diameter. That's a lot of acreage.

For that reason, I call the Eggmobile a land extensive system. The pastured broiler shelters and the Millennium Feathernet are all land intensive systems. (For complete diagrams and specifications, see my book *Polyface Designs*.) The differences between extensive and intensive are critical for small holdings. Chickens, like all animals, love routine. Cows pick favorite lounge spots and even though another area may be clean and sanitary, they will go lounge in their manure where they lounged yesterday. Animals are creatures of habit.

The only reason the Eggmobile works for us at Polyface is because we can move it far enough each day or two to keep the chickens unfamiliar with their surroundings. The only thing familiar to them is the Eggmobile. In fact, if we move it just 50 feet during the day, they will not be able to find it in the evening when they want to go to sleep. They'll huddle down in the middle of the field where they exited in the morning rather than going 50 feet over to where we moved it at midday. Don't ask me why. They readily range out 200 yards, but can't find their way home from 20 yards if you move the Eggmobile after their morning exit.

This homing instinct and desire for routine creates a problem with unrestricted free range chickens on small properties. On a five acre homestead, for example, nothing is more than 200 yards away from anything else. That means the chickens can never be de-familiarized with anything. I know that's not a word, but I think it conveys best what I'm trying to get across. The chickens will find a comfortable area and set up housekeeping there.

And because the property is small, you can never make them stop going back to that spot. Certainly you can have a stationary coop that they return to each evening but you'll probably have an Easter egg hunt each day. On small acreages, I suggest letting them out a day or two a week only. That way they won't get too familiar with the lawn mower handle and the porch swing and will be more disciplined to return to their designated quarters. I recommend this even if you use a portable coop. Limit their access to the whole property and they'll be less likely to settle where you don't want them.

In this homing and ranging regard, all the fowl are similar: turkeys, ducks, geese. As they get more wild, like guineas or pheasants, this homing and ranging tension becomes more acute. These wilder fowl range farther and can be hard to train to return to a certain place. They may pick a place, like a favorite tree for roosting, and try as you might, you can't change their mind. The more they can establish their own routine due to being hard to control, the more difficult it will be to get them to lay their eggs or sleep in a certain place.

Lastly, nothing disturbs the landscape like a pig. For that reason we don't run pigs in our nice pastures because we don't want divots everywhere. If that is not a problem, or if you have a disc or harrow to rake out the divots, then by all means add pigs. Just remember that timing is everything on this disturbance. If you create a moonscape with the pigs, you'll have hard soil and have to probably till it in order to plant anything. When I began experimenting with pigs on pasture, I built a 12 ft. X 20 ft. enclosure named the Tenderloin Taxi.

I simply used locust poles for a base, connected with angle iron and bolts at the corners. I cut some large diameter pipe and welded the arced pieces to some short pipe uprights. Lag screws in the concave butt pieces fastened the uprights to the locust poles.

I then wired hog panel to the uprights, leaving one little spot for a gate, and put some boards and shade cloth across a corner. I put four hogs in there and moved it once or twice a day with the tractor.

I wanted to see how many square feet a pig would tear up in a day given different conditions. If the ground was extremely dry, they didn't dig much. But if it was wet, they'd tear it up rapidly. As I moved the Tenderloin Taxi along, the trailing pole bottom would help even out the divots, but certainly not completely. Sometimes I'd fling in some small grain, let the pigs tread it in, and then I'd have green squares of annuals where the pigs had been.

What I'm describing here, of course, is disturbance on steroids, but I'm trying to lay out the full spectrum of disturbances, from least to most, in order to offer options. And to caution you about what can change if overnight you receive a one inch rain. You can't know what you haven't heard. We only used the Tenderloin Taxi for a couple of years because we didn't want our pastures divoted and we didn't have a disk or harrow. But if you're trying to really tear an area to pieces or prepare a garden spot, pigs will do the job.

If the ground is too wet, they'll turn it into bricks. If it's too dry, they won't do much. It has to be moist but not wet. If you pick up a handful of soil, you want it damp enough to squeeze into a ball, but not damp enough to shed water when you squeeze it. When the moisture is just right, pigs are a tillage tool from heaven.

For several years we ran pigs through our garden in little 6 ft. X 8 ft. wooden gate boxes hinged on the corners. I could pick up one side enough to move it forward a couple of feet, turning the box into a parallelogram. Then I'd walk over to the other side and move it forward two feet to square, then two more feet. Repeating this procedure, I could walk the box forward through the garden.

With two pigs in there, we could easily till two box spots a day, or 96 square feet. The problem was that when the time was right to plant garden, we needed a lot more than that to plant every day. We needed a bunch tilled at one time and then nothing for awhile. The window of opportunity was so small it worked better just to use a broadfork or tiller and we eventually abandoned these pig cultivators. Today we're close to no-till gardening, just maintaining a heavy mulch year-round. But this is not a book about gardening.

We did use these boxes for a few years in our hoop houses in the winter with the laying hens to till the deep bedding and keep it from capping. It worked well but eventually we found it easier to throw whole grains on the floor to stimulate scratching. Pig power is quite amazing if you can harness it at the right time to do the right thing. Just remember that if you have dirt floors they will dig to China and whatever field they're in will be uneven and torn up after they pass through. If all you're running is herbivores in that pasture and you don't plan to mow it for hay, the unevenness isn't a big problem. But if you're using electric netting or pulling chicken shelters, you'll curse the day you put pigs on the field.

So far, we've looked at renovating worn-out pastures. All these regimens are focused on awakening a latent seedbed to move vegetative succession forward. So far, I have not encouraged you to plant anything. It's all about resurrecting the legacy seed bank.

Now we'll shift gears. What if the field you're working with has been a crop field and you want to establish a pasture in corn stubble or soybean residue? To be sure, given enough time, you can use all the principles above to create a pasture. But it'll be a decade before you enjoy a productive conversion. Most of us don't want to dither that long.

For those of you dying to plant a seed, here's your opportunity. Remember that nature does not place seed into the ground as much as it drops seed on the ground and lets gravity and time work it in. To sprout, a seed needs moisture and to get established, it needs what is called seed-soil contact. A seed that sprouts half an inch away from the soil will quickly die.

Planting protocols differ from region to region so I hesitate to give any blanket recipe. Just remember you need seed-soil contact and moisture. If you're planting into crop residue, you can't just fling the seed and hope for the best. Even on bare ground it'll perish unless you get it pressed onto the ground. If you plant and then don't get any rain for two weeks it'll perish as well, especially if it doesn't have good soil contact. These are universal principles.

Here is a poor-boy technique we've used throughout the years to good effect. A hand-held cyclone seeder is cheap and on a windless day will spin seed out over a 12 foot swath. We cut some big (12 feet tall or more) bushes or small trees to create a nice clump, tightly chain their butts together, and then chain the whole lot to the back of a pickup truck. One person can sit on the tailgate with the hand seeder spinning off the seed. The tree tops dragging behind disturb the soil and crop residue, lightly covering seed and in general stirring up everything enough to cover the seed.

The nice thing about this technique is that you don't need an implement to cut through the crop trash. Also, the plants don't come up in rows; they come up randomly more like a carpet. Soil conditions must be just right in order to do this. The soil can't be too dry or it's too hard to get any scraping or movement. It can't be too wet either. It needs to be a tad dry on top with lots of moisture directly beneath. You definitely want to stir up some dust because that shows you're getting some soil movement.

For pasture, my favorite seed mix is a cocktail of at least two legumes, four grasses, and at least one small grain as a nurse crop. The small grain comes on much faster than the legumes and grasses, offering some quick ground cover to both shade the soil and hold it in place until the grass plants get some size. Generally the preferred nurse crop is barley or oats. Ask your seed dealer for pounds per acre and pad it a little bit. The technique I've described here is less precise than a precision drill; for insurance, plant about 20 percent heavier than recommended. That will still be much cheaper than hiring a drill.

If you're planting a larger field (anything more than 3 acres) you might want to borrow a 3-point hitch mounted cone-shaped hopper fertilizer spreader. Those work great for seeding and you can pull your tree top soil disturber behind the tractor; this is a one pass planting system. If you are averse to gossip and rumors, you might want to do this when no neighbors are watching because I guarantee you dragging a bundle of tree tops and bushes across a field definitely attracts attention and starts tongues wagging.

On a brand new planting like this, be gentle with the grazing. It'll take up to three years to attain full strength. No pounding, close grazing, or high impact early on. Wait until the forage gets a foot high before lightly grazing it the first time. Let it get established before intensifying the grazing regimen.

Before leaving this topic, I want to introduce you to pasture cropping. Developed by Australian Colin Seis, pasture cropping uses livestock to temporarily weaken pasture in order to grow an annual. You might want to grow some sudex or even forage corn for a highly palatable, high energy summer heat-loving option.

The protocol requires an initial grazing, then planting with a no-till drill, then another grazing right before the annuals break ground. For example, we've done this in May with corn, cowpeas, millet, and sudex. We graze about May 10, wait about a week, and then plant the annual. If the annual breaks ground a week after planting, we'd graze again May 24. The grass is still fairly short and just recuperating from the earlier graze.

To be sure, the cows aren't happy on this second grazing, but you're doing a one day pain to get gain later in the season.

This close second graze weakens the perennial sod enough to enable the annual to sprout and jump ahead of it. The annual soon shades out the perennial and grows apace. When you graze the annual, the shaded perennial grows quickly. This technique eliminates tillage (except for the sod drill) and keeps vegetation on the ground at all times.

Seis pioneered pasture cropping with sheep and small grain, but we've duplicated it with cows and sudex or corn and cowpeas. In our experience, the larger seeds work better than smaller ones. Another permutation is to do this with a cool-hardy small grain in the fall to create palatable green forage in early winter and early spring. Most of the cool-hardy small grains grow well at temperatures a dozen degrees below what is required for grass to grow.

Argentina grazing aficionado Anibal Pordamingo calls this a forage chain. By linking together native forages with cool season and warm season annuals, you can keep something high in sugars in front of the herbivores year round or almost year round. Speaking from experience, let me caution: do not attempt this unless your field is extremely fertile. Annuals require large amounts of nutrients; this procedure will not build your fertility.

It can certainly use excess fertility and maintain fertility, but a forage chain cannot build fertility. Unless and until you can get proper growth on the annual, don't plant it. Wait until you have good fertility, or lay down a bunch of compost when you plant it. You'll be disappointed if you try planting annuals into low fertility soils.

Extending your grazing season either into the hot dry summer or cool wet winter with strategic annuals is akin to using cold frames and tall tunnels to extend vegetable growing seasons. While pasture cropping is in its infancy right now, I predict it will find more practitioners as expertise in this technique increases.

Since you need moisture to guarantee germination, be sure to try this only when you have good moisture and dependable rain. Obviously this is another opportunity to use irrigation as an insurance policy. With irrigation, you could plant any time the weather is warm enough and know that you'll get germination. Some simple irrigation can pay big dividends when leveraged strategically on a season-extending forage growth plan.

Establishing pastures is an art as well as a science. I hope this discussion shows the many nuances wrapped up in pasture development and overall forage enhancement. You'll never tire of tweaking your pastures; that's your lifetime of fun. You'll gain insight on every experiment. As forage cocktail guru Gabe Brown advises, the main thing is to do something different. Amen.

Chapter 11

Converting Woods to Pasture

Livestock needs pasture. Many homesteads are forest rich and pasture poor. A five acre place with one acre of pasture and four acres of woods severely limits livestock options. While you can run omnivores through the woods--especially pigs--the most desirable habitat for all the domestic livestock is pasture.

The wonderful thing about good pasture management is that you don't sacrifice ecological equity making the conversion. In many industrial and orthodox conversions, that is not the case. Converting it for corn production or any annual cropping, for example, does not preserve the soil building and carbon sequestering capacity of the forest.

Don't let anybody throw a guilt trip on you for converting some woods to a productive pasture. To be sure, sometimes the reverse needs to happen. On our own property, we reforested about 60 acres in the first few years. Our general rule of thumb is that if the terrain is too steep to feel comfortable driving on, it should be in trees.

The homestead property you buy may suffer from years of neglect. It may have been vacant for several years before you arrived on the scene. Especially in temperate areas with rainfall above 20 inches per year, land abandonment sets a succession course toward forest. Certainly east of the Mississippi River, all pasture is borrowed from the forest. If you quit mowing your lawn, it will eventually turn into forest, which is the climax ecosystem east of the Mississippi. River.

A property lacking human intervention will eventually turn into forest in these temperate climates. The vegetative succession depends on fertility. If fertility is low, the first species will often be briars and impenetrable brambles followed by pioneer saplings and then the more long-lived trees. In our area, these pioneer saplings are often conifers, which require full sunlight, can grow in compacted soil, and provide a nurse partner for the long-lived oaks, maples, and poplars.

If the soil is fairly fertile, we might see walnuts and locusts sprout up within a year or two of abandonment. Squirrels plant lots of walnuts. These saplings can grow bigger than you'd want to drive over with a tractor within three years. I call this encroachment. It can occur in abandoned fields but it can also occur along field edges next to the woods.

That's the zone of most rapid forest colonization because the tree environment is already proximate and seeds are abundant. Trees along a field edge always lean out into the field seeking sunlight. As these edge trees reach over, their lower branches impede access underneath, as anybody who has been bonked on the head with a branch when trying to ease up close with a tractor can attest.

If these reaching edge trees are not aggressively culled out, the woods will gradually encroach into the field. One year you lose a yard and the next another yard and before you know it, you've lost 20 yards along the field perimeter. In 250 yards of perimeter, that amounts to an acre of pasture. Cleaning field edges from limbs and encroachment is one of the normal maintenance chores on any property. Just doing that can often provide the season's firewood even on an extremely small homestead.

Obviously if the property laid fallow for several years prior to your arrival, this field edge encroachment can be severe, with trees up to a foot in diameter occupying what was one time a legacy field edge. Whether the tree issue is this edge encroachment or overall field encroachment or you just want to make a full-on conversion from a patch of woods to pasture, you can create beautiful and functional fields with simple tools.

I'll make an assumption here that you aren't planning to grow annual crops. If you're going to do that, you'll need to get rid of the stumps. But for any other application except small broiler shelters, low stumps are okay. On our farm, we've done numerous conversions over the years totaling more than 20 acres. We've never done more than 1.5 acres in a year and we've not used any excavation equipment except on a one acre spot crossed with gullies. We wanted to push them shut in order to smooth it out a bit, but otherwise, we've done all our conversions with a chainsaw and tractor.

The biggest and most rapid conversion was a six-acre field that had not progressed to full forest stage. By the time we tackled it, the primary vegetative cover was a horrible and massive thorn bush that grew up to 15 feet high and sported three-inch thorns. We called them hypodermic thorns. Because they were bushes, gaining access to the trunk required body contortions to stay away from the low-lying thorny branches. They were much too big to push over with our 40-horsepower tractor.

The worst part of the project was that because they were almost round shaped, many did not easily express their lean direction. To protect me from getting impaled (I ran the chainsaw), Dad flipped a rope up on the bushes high enough to catch some branches. He pulled on the rope to help the bush fall away from me as I cut the base. I'm belaboring the story a little bit to encourage you to tackle things that seem intimidating at first.

I remember when we made the decision to go start the project, firing up the chainsaw, and attacking that first bush. That six acres looked like an insurmountable project; how could we ever, cutting just one at a time, open up the whole field? But you know the old saw about how do you eat an elephant? One bite at a time. A close corollary could be how do you re-open an encroached field? One tree or shrub at a time.

After working for an hour, we had a nice little opening. After another hour, we'd conquered nearly a quarter acre. Just 23 more quarter acres to go. And that's exactly what we did. We used the tractor to push the cut bushes into piles. As soon as we got half an acre opened, our adrenalin kicked in because we could see progress. It was a pasture development snowball. Over the course of a couple of months, we finished the field and had multiple stacks of bushes waiting to be burned.

We waited until a damp, still day and began lighting fires. Wow, was that fun. That's when the field really started to look like a field again. That project included a trip to the Emergency Room to cut a thorn out of my head. I bought a hard hat on the way home from the hospital and have religiously worn it since then; if I'm using the chainsaw, I wear a hard hat. Let this be a lesson to you, dear friends.

Obviously small stumps peppered that field for several years, but by about year six they deteriorated enough that we could kick them out with our feet. Today, that is one of our most productive and nicest hay fields. It just needed some care and attention; abundance required caress.

Not all conversions are as complete as that one. It had no large trees in it and no locusts; all the stumps rotted quickly and completely. I cut the stumps as low as I could get with the chainsaw and not hit the ground. That's not too hard to do on smaller material, but on big trees, it's much more difficult.

If you walk through a cut-over forest, you'll typically see stumps about a foot high. I know that as I started this discussion about conversion, many of you had that image in your mind and wondered how in the world a decent pasture could be created out of a field of stumps like that. Cutting stumps a foot high, which is still much lower than the waist height of yesteryear's crosscut saw era, makes sense for a couple of reasons.

First, stooping over with the chainsaw any lower is awkward and tiring. You can't get good leverage and the saw cuts down around your feet--not a good thing when felling and paying attention to everything else going on. Secondly, of course, it keeps the saw high enough from rocks and dirt to protect the chain from inadvertently hitting something abrasive. Third, a tree butt swells as it enters the ground. Cutting the butt just above that swell allows for more precise felling cuts and reduces the total width to cut (less diameter).

Those are all good reasons to cut a tree about a foot above the ground. I employ the same practice. But then I have another protocol. Once the tree is down, I make a second cut as close to the ground as I can without hitting dirt and rocks. Now the pressure of the felling is gone and I can concentrate fully on simply cutting horizontally through the left-over stump. I can hear some of you expert sawyers lamenting that this is an unnecessary laborious step, but I can tell you that safety dictates taking this two-stage approach in order to get the look you want.

With the pressure of the felling gone, I can kneel on the ground if I want to get comfortable and increase leverage on this horizontal cut. You would never want to kneel and get comfortable that close to the tree while you're felling. You want to stay nimble, up where you can move quickly in case things don't go according to plan. It often takes awhile to get through the swelled bottom of a big stump. With my 20-inch bar, I can get clear through about 36 inches. Sometimes even after going all the way around I can't get enough of the middle to bump clean with the tractor, but not often. If you have a friend with a longer bar, or want to rent one, that's fine. In our area, this happens seldom enough that on those stumps, I just leave them. One or two of those per acre is not a pasture-killer. That massive stump can remain as a monument to the grandchildren of what once was.

Once I get the slab off, I scoot it to the side to do the final sculpting. With the snout of the saw, I cut all the way around the edge of the stump on a diagonal, making a nice beveled edge. You can go either forward or backwards with the saw; I prefer going backwards because I feel like I have a bit more control. Usually I can't go all the way around in one continuous cut; normally I'll take three or four slices to completely cut around the stump. Normally one side of the stump will be a little higher than another so a slice will be deeper on one side. The point is you're trimming up that stump to round the edges.

I can't tell you what that does to a forest harvesting project. I do this on areas even when I'm planning to let the forest grow back. Why? It's pretty and functional. With these beveled edges, you can drive over the stumps with machinery (not hay equipment). The beveled edge makes the stumps gentle to roll over, like an inverted saucer. For photos of my technique see pages 84-85.

The biggest reason to bevel the edges is to allow overpass with a bush hog safely. If you leave the edges vertical and a rotary brush mower hits them, you can easily break a blade. But with the beveled edge, the blade rides up over it gently or at least you can hear it clang before it goes clunk. Been there, done that.

Most rotary brush mowers come with stump jumper plates. These saucer-shaped protectors below the blades keep the vertical drive shaft from snagging on things. Stump jumper design revolutionized the ruggedness of rotary brush mowers.

Once you've cleaned the area, you have a new spot of soil with no protection. The number one issue in this conversion is getting it vegetated in grass immediately. If you delay getting forage planted even one season, it'll be the worst weed and bramble nightmare you can imagine. What arrests the aggressive re-capture by the forest ecosystem is a proactive aggressive line in the sand that says "you are now a pasture, not a forest."

In our experience, no additional land preparation prior to planting is necessary after we pull out with our last load of firewood or saw timber. After driving over the area with tractor and trailer, any small vegetation we didn't cut with the chain saw is crushed into the soil. If you have significant small woody species still upright when you pull out, I'd recommend going over it with a rotary brush cutter to make sure you minimize competition for your new seedlings.

When you do your clearing work has an important bearing on the initial seed planting. If you start on the project in late summer and work on it through the winter, you might have a few weeds that come on before winter, but it won't be many. If you do your clearing in the spring and wait to seed it the following spring, you'll have a mess. We prefer to do our woods work in the winter because that's when we have time and the snakes are dormant. The poison ivy leaves die off, too.

Whatever we finish by late winter gets the spring seeding treatment. I cannot emphasize strongly enough the need to get a good seedbed growing immediately coming into the growing season. Nature does not like nakedness, and your newly naked soil will get covered with something, like it or not. You might as well get it covered with something you like.

I've already talked about seeding crop fields using a cyclone seeder and a drag made out of tree tops. This is a perfect technique for the newly cleared, bevel stumped place. You can't pull a typical seeder through the area due to the stumps. But you do need to disturb the soil enough to get some seed/soil contact. This technique works in any and all terrains as long as you do it in the right season: soil damp, disturbable, and warm.

I seeded one conversion project with a cyclone seeder and then cut some brush, tied the bumble to my waist, and walked along, dragging it like a bridal train. It worked perfectly. I did the same thing with an ATV once. If the soil is right, you can get disturbance with a bundle of brush dragged along.

An alternative to planting seeds is the hay feeding method. Pick the kind of hay you use carefully. It can't be nice second-cutting. It needs to be old-cut seedy without mold. If it's moldy, it heated and cooked the seeds. It can certainly be spoiled around the edges, like a 2-year old round bale.

Throw out this hay liberally, about twice as much as the animals will eat. They'll pick through and eat the cream but stomp and spread the rest onto the ground. This gives the advantage of mulching the soil in your new field and buys time for seeds to germinate prior to weeds growing. Don't be stingy; the whole goal here is to have enough hay left to cover the soil after the animals have eaten what they want. This is also our favorite technique on new pond dams to get them stabilized and seeded after excavation work.

Of course, if you really want insurance, spread the seed and then feed or spread hay on top. In this case, you won't need to disturb the soil; the hay mulch will lock in enough moisture to ensure germination and provide enough settling onto the soil for good rooting. If the weather suddenly turns dry, some water sprinkles would be helpful, but normally if you do all this late in the winter or very early in the spring, moisture will be adequate.

Your seed planting will come on gently. Obviously spreading compost will help the plants become robust sooner. Don't overgraze the new planting; be gentle the first year and watch your pasture lengthen and stabilize. If for whatever reason you have some holes, you can always go back and sow those, like a spot treatment. By all means keep chickens off this new pasture for the first year. Chickens can tear new seedlings out of the ground in a heartbeat.

We've used these techniques for many years in numerous applications. Low-cut stumps do not impede any animals.

They don't even impede electrified netting, Eggmobiles, or a Millennium Feathernet. As long as your shelter has enough clearance, it'll slide right over these stumps. Beveling makes all the difference.

Finally, if for whatever reason you get a thin stand or weeds and brambles begin germinating, don't be slow to mow with a brush mower. If you set it high, it'll go right over the stumps. Often the place where brambles and weeds germinate most aggressively is right around the stumps because that mound, however slight, impedes seed placement, soil disturbance, and mulching. This is why you must have the stumps low enough to mow and drive over.

Many newbie homesteaders are intimidated by this conversion recipe due to elementary chainsaw skills. I encourage everyone to develop mastery with a chainsaw; it's probably the most significant innovation in the food and fiber sector except for electric fence. Take an online safety course or sign up for one of the courses offered through your state's Department of Forestry. Everyone who wants to caretake a piece of property should aspire to competence with a chainsaw. That said, big trees are still a bit intimidating and scary. That's okay; find a friend more skilled to do the most difficult cutting.

I advise not using a commercial logger for the project. First, unless a project is sized significantly, a logger won't move his equipment to your site. And by significant, I mean more than a dozen acres. Second, you'll never be able to clean up the site to the above specifications in time to get it seeded. You'll be several years getting it cleaned up and by that time you'll have 12 foot saplings again and the place will be covered in impenetrable blackberries 10 feet high.

If you actually have a dozen acres to do, I suggest pecking away at it for several years. In our experience, a one saw and farm tractor operation can do about an acre a winter. That's a pace that lets you sell the firewood, chip or burn branches, mill the saw logs on your or a friend's band sawmill, and be assured you'll get things pretty and seeded in time. These kinds of projects have a much higher success rate if you go at them methodically, meticulously, and micro-scale.

If you end up asking a friend to do this work for you, make him read this chapter so he understands what you're trying to do. I don't know anybody on the planet who bevels the edges of second-cut ground level stumps. I also don't know anybody who leaves a forestry project looking as manicured as mine. Just sayin'.

To speed up stump decomposition, the old folklore recipe is to bore a couple of holes in the top and pour in buttermilk. Many stumps disintegrate in less than a decade. The only ones that take a long time are locusts and osage orange--fence post material. If you want an instant clean slate, you can rent a stump grinder.

I never, ever recommend doing a conversion with excavation equipment. That's like trying to cut your fingernails with a hammer and chisel. It's expensive, tears up the soil, often scraping off the topsoil, and leaves the beautiful biomass in an unusable form. Even if you can salvage some of the firewood out of the piled debris, it'll be full of dirt and rocks, dulling your chainsaw and aggravating your spirit. No, leave the bulldozer out of this project.

Some of our prettiest, most productive fields today were solid woods a few years ago. You can make the conversion. Just start. Don't try to do too much in one go; pace the project. Half an acre done well is far better than two acres done poorly. Enjoy the process and the transformation as you bless your homestead with some more pasture.

Chapter 12

Management Intensive Grazing

When Jim Gerrish switched the phrase "intensive grazing management" to "management intensive grazing" it represented an explanatory cataclysmic shift in the grass farming movement. Early on, practitioners emphasized the grazing intensity as the signature characteristic of the movement. But the Gerrish switch emphasized the management aspect and that truly was a game changer.

All of us commercial practitioners knew we were putting way more animals per acre on our pastures and looked at that as grazing intensity. It was, for sure, but the underlying ***why*** for putting on way more animals per acre was predicated on proper management considerations. Gerrish's epiphany reverberated throughout the grass farming community and all of us smacked our heads with an "of course, why didn't I think of that?"

Strict grammarians (as a college English major, I are one) would say that the phrase should be hyphenated: management-intensive grazing. But Jim didn't, and neither did Allan Nation, who popularized it through the iconic pages of the *Stockman Grass Farmer* magazine. Far be it from me to parse that oversight; I'll use it without the hyphen too. Now that you're thoroughly grounded in the background of the phrase, what does it mean?

It means that ecologically authentic grazing starts with intensifying our management. This is critical because my sense is that only 5 percent of all domestic grazing animals in the U.S.--perhaps even the world--are intensively managed. They are put in places and left there for extended periods of time. Even movement in nomadic cultures seldom considers all the management nuances necessary to increase forage and maintain ecological progress. Most of us are familiar with the tragedy of the commons in these regions.

In fact, so common is grazing mismanagement that the word grazing is vilified in many environmental circles. That desertification, ecological destruction, and poverty all coincide with grazing gives pause to any domestic livestock scheme. In day-to-day conversation, we graziers feel compelled to apologize for what we do. Or at least to apologize for the conventional practitioner as we distance ourselves from "those farmers and ranchers."

Grazing is demonized enough now world wide that none of us who manage grazing animals can affiliate with that fraternity without caveats, explanations, and separation. These days, in normal conversation, I never say, "I'm a farmer who raises livestock." Such a statement invokes wrath from animal welfare, climate change, vegans, fake meat and skinny jean health cultists. I have to say instead "I'm a pastured livestock producer." All my terminology and explanations distance me from conventional practice.

I'm belaboring this because it is the single biggest hurdle for small holder livestock producers. The most overgrazed and wretched pastures I've seen in my travels have been on homesteads. Why? Because either small holders don't know how to apply management intensive grazing principles or think that they can fudge the rules because their outfit is small. I want homesteaders to teach commercial farmers how to utilize management intensive grazing. I understand that if you have

2,000 acres, the idea of moving the herd every day through a series of controlled paddocks could be daunting. After all, you have fencing, water, and logistics to figure out.

But just like it's easier to turn a speedboat than an aircraft carrier, it's far easier to implement good grazing management on a small acreage than a large one. Be encouraged by this and commit to doing it well. Here we go.

The foundational principle is the grass growth curve. Like almost all biological beings, grass begins to grow slow, then speeds up, and then goes into senescence. A human baby starts to grow (will she ever get out of diapers?) and then hits stride, accelerating through the juvenile growth spurt, and then slows down and stops by roughly 20 years old.

Plants go through the same phases: infant, juvenile, and maturity. The engine that drives all this activity is the sun. Through photosynthesis the plant metabolizes sunbeams (yes, I know it's sunlight, but sunbeam sounds more like fantasy and dreams) into physical structure. I've never quit being awestruck at the miracle of that process. That something as non-physical as sunbeams can be converted to something fungible, measurable, weighable, and structural never ceases to amaze me.

Because of this growth pattern, the plant has times of relatively high efficiency in converting sunbeams into structure and periods of relatively low efficiency. Obviously if the plant reaches maturity and then senescence, it turns brown and quits taking in any sunbeams altogether. At that point, sunbeams are being wasted because they are striking vegetation that can't metabolize them into something useful.

On the other end of the spectrum, if the leaf is barely visible due to seasonal dormancy (winter) or overgrazing, the plant has no solar panel to metabolize sunbeams. The green leaf determines the efficiency at any given point in time. If an animal prunes off the top of a plant, the plant relies on stored carbohydrates in the crown and root to provide the energy in sending forth new leaves and new shoots. A plant either relies on this carbohydrate savings account in its core and root system or its leaves to provide operational energy at any point in time.

Wouldn't it be nice if it were all as simple as moving the animals along without ever severely pruning the grass in order to preserve enough leaf area to never have the plant go back to slow growth infancy stage? It's not that easy due to several factors. The first is that palatability varies from plant to plant and from stage to stage. Animals always eat dessert first. Always.

Animals prune the most succulent and satiating plants first. And they often graze the most desirable plants short before considering less enjoyable neighbors. This is why you can have overgrazing and undergrazing at the same time on the same square foot of pasture. I call the pasture my salad bar and will not repeat information in my book *Salad Bar Beef*, but suffice it to say that palatability differences greatly affect species composition.

If a desirable plant receives constant pruning while its neighbor grows tall and makes seed, over time the hammered plant will weaken and die while the ungrazed plant becomes more robust and proliferates. Clumped up animals will definitely eat far more variety at a more reckless pace than non-clumped animals. This leads us to the concept of mobbing, which is hard to get with a handful of animals; hence, our discussion about disturbance.

Animals don't just eat; they also tromp and lie down. What they lie on, walk on, and poop on quickly moves to the undesirable status. The faster our animals ingest what they want and quit walking, the higher the percentage, or utilization rate, of the available forage they will ingest. Animals walking willy nilly around a pasture actually soil much of the available forage. You want them to eat and then chew their cud. An herbivore does not make meat or milk when grazing; only when ruminating.

Cows are union members; they only work 8 hours a day. They need 16 hours to chew their cud. If the pasture is overgrazed, they spend too much time trying to get something to eat. A cow prefers blades around 8 inches long. A sward 12 inches or more tall provides extremely efficient fill-up for the grazer. If the sward is too short, the animal spends an inordinate amount of time trying to fill up with tiny, short blades. It would be like you trying to eat supper with a toothpick instead of a fork.

In the perfect world, you only offer the animals one day's worth of forage at a time. That reduces waste and encourages the greatest ingestion variety. On our farm, our cows routinely eat thistles and sour dock and other weeds they would normally not eat. The small paddock each day encourages a more even graze. In a fast growth period like late spring, healthy forage in fertile soil often grows back an inch a day. An herbivore can certainly eat a plant with an inch of blade on it. That's like a candy bar to the animal.

But it's deadly to the plant because that new shoot required stored carbohydrates to come forth. If the plant must dip into its savings for yet another shoot, it can't replace that lost energy as quickly as if the first shoot had never been nipped. Andre Voisin, the French godfather of grazing management, called this "the law of the second bite." Anything we can do to prevent that from happening yields big dividends in increased plant production and robustness.

Plants don't grow at the same rate year round, which means the wait period until it's ready to graze again varies widely. That is why we never, never, never make permanent paddocks. We do as little permanent fencing as possible and depend on portable fencing to define the daily paddock. Voisin called this "rational" grazing because you're actually rationing out the forage in the field just like you would bales of hay.

The portable fencing is the brake, accelerator, and steering wheel on that four-legged pruning sauerkraut vat. We'll talk a lot more about fencing in another chapter, but for now you need to understand why it's necessary and must be mobile. If you graze through a field four times in a year, each of those four times will require different paddock sizes because not only will the forage be of different volume and quality, but also the animals will increase or diminish.

You might sell one. One might have a calf. A lactating cow eats 30 percent more than a dry cow. Same animal but different needs. Flexibility is the key. Just imagine your animals like a lawnmower that you're steering around your homestead, placing them where they need to be one day at a time. The result of this management is a mosaic of different-staged forage. In general, you want a third of your pasture acreage too young to graze, a third of it ready or about ready to graze, and a third of it either waiting or just grazed.

We already discussed letting some areas blow out to encourage seed shatter and diversity. In those cases, of course, the animals won't eat but perhaps half of the forage; they'll stomp the rest onto the ground and all that dry lignified cellulose feeds the soil microbes and mulches the surface to keep it moist. Sometimes you'll want to harvest lightly and other times you'll want to really sheer it off. Your animal needs will vary as well. A growing calf needs more protein than a finishing calf, for example.

Young, tender grasses have more protein than more mature grasses. Besides my book on grass finished beef production, plenty of others can provide you lots of nuances about managed grazing. I won't repeat that here; this is just a cursory introduction and an explanation of objective. You want to prohibit the second bite on a recuperating plant until it has matured enough to replace all the carbohydrates in its savings account that were expended in sending forth the new shoot. That varies from plant to plant, but generally it's in the early boot stage before you can see any seed heads coming out.

Another general rule is fast growth, fast movement. Slow growth, slow movement. In other words, slow growth requires longer rest periods, which means you're increasing the days in the rotation. You do that by decreasing paddock size. That is why much of our planning is geared toward stockpiling a bank of forage out ahead of the animals so that in cold or dry conditions we're not suddenly out of grass. As you gain experience, you can walk out each week and take an inventory of what you have ahead. That way you won't get caught short and can make decisions with plenty of lead time.

Voisin called running ahead of your grass growth "untoward progression." This is extremely common due to different growth rates. If you don't let the grass go through that accelerated juvenile growth phase, the overall volume in your forage inventory decreases. The animals go through their paddocks faster with the lower volume and the next thing you know, you're grazing infant grass and nothing of any volume is out in front. Suddenly you're into feeding hay. You have to match the stock to the acreage to the growth cycle. And yes, hay is a ballast that you use seasonally when you have no other options.

Routinely people who adopt this management protocol reduce overall hay feeding by half or more. That's a significant savings in time and money. The reason is because you can bank a standing inventory ahead and you can actually measure how many days of grazing you have out in front of you. Without this level of management, you're flying by the seat of your pants. In a continuous grazed field, nobody really knows what's out there.

A plant maintains biomass bilateral symmetry between the above ground and below ground portions. Plants that never get tall have roots to match. If you routinely let the forage get taller and even go to seed once in awhile, the plants will establish large, deep roots. That keeps things green through a drought and creates overall vibrant activity in the soil, which warms it and awakens it earlier in the spring. All of this amounts to season extension in dry or cold times.

Before we go into the mechanics of this, let me give one more principle: change it up. You'll be tempted to run the same rotation every year because, for example, a south slope will always green up faster than a north slope. You're feeding hay and desperate to get out on the grass. You see that greening south slope and you can't help but turn the animals out on it. If you do the same thing to the same area at the same time every year, you will deteriorate your sward.

If you started at paddock A last year, start at paddock E this year and paddock G next year. That way you're touching areas at different times of the year from year to year. That allows different species to reach phenotypical expression. The plants don't look at a calendar; they grow when conditions are right. By changing up your year-to-year rotation you'll diversify the sward and protect yourself from overgrazing the succulent plants.

Okay, let's jump into the mechanics. You understand the grass growth curve, the law of the second bite, the imperative to clump, prune, and rest. Now the 64 million dollar question is this: "How do I know how much to give them?" That, my dear grazier, is a question no one can answer but you on your place with your animals in your situation.

But here is how you figure it. I'm going to be loose with some figures and hope not to irritate my scientist and engineer readers too much, but round numbers are easier to do in your head. An acre is roughly 5,000 square yards. That's about the size of a football field. The formula for figuring how much area is based on Animal Units per Acre (written AU/A). Of course, you have a time element: a day. Because it's easy to say and we have cows on our farm, we use Cow-Days per acre rather than animal units. Here is the formula: **cows X days divided by acres.** If you can't remember that, you can buy a T-shirt from our Polyface gift shop with the formula prominently emblazoned on the back. Ha!

Acres are constant. Days are constant. How do we make animals constant? After all, a week-old calf is certainly not equivalent to its mother when figuring how much grass it will eat. This is why whatever you want to call your animals, you have to arrive at a consistent measure. On our farm, we use cow equivalents. Our cows are relatively small (1,000-1,100 pounds) so that poundage becomes our standard measure. Just like inches for a carpenter and bushels to an orchardist, the cow equivalent is our standard.

If we have two cows and two one-month old calves, we have 2.2 cow equivalents. At weaning, these four animals will become 3 cow equivalents because the calves will have grown to half the size of a cow. A mature bull is 1.5 cow equivalents. A mature sheep is 1/7th of a cow equivalent and so on. Every flock or herd carries a cow equivalent designation. If we know the cow equivalents and the area we gave them for a day, we can quickly determine the cow-days per acre for the area we gave them that day.

If I have 10 cow equivalents and I give them a tenth of an acre, that's 10 cows X 1 day divided by a tenth (X 10) equals 100 cow-days per acre. If I have 5 cow equivalents and I give them a tenth of an acre, that's 5 cows X 1 day divided by a tenth (X 10) equals 50 cow-days per acre.

If you don't know your cow-days per acre, you have no basis for adjustment. If you feel like they were short and they look or sound grumpy and you decide to give them 10 percent more than yesterday, if you don't know how many square yards you gave them yesterday how do you know what 10 percent more is? Although I said square yards, we work in acres or increments of acres. Goodness, I've given 100 cows a third of an acre a day when the forage is extremely lush and tall.

A third of an acre is roughly 1,600 square yards. If I'm going to give them 10 percent more next move, I'll give them another 160 square yards. If I want to give them 10 percent fewer,

I'll reduce the area by 160. If you don't do the math, you can't do the management. Eyeballing doesn't cut it. Your eyes will deceive you every time. By doing the math and keeping records, you will soon develop what Allan Nation called "the grazier's eye."

That's the ability to look at a sward and call it 30 cow-day or 50 cow-day or 10 cow-day. If you have 5 cow equivalents and you have 50 cow-day grass, you'll give them a tenth of an acre for the day. You thought you weren't ever going to use your 8th grade algebra, didn't you? That's all this is. If your grass isn't very good and is only 20 cow-day, your 5 cow equivalents will need a quarter of an acre. What does a quarter of an acre look like? If an acre is 5,000 square yards, a quarter of it would be 1,250 square yards. Well, what does that look like?

Once at a seminar I was going through this and when I asked this question, a guy on the front row said "1 yard wide and 1,250 yards long." Wise guy. That would be a fairly inconvenient paddock, methinks. Let's be practical. What does a 1,250 square yard paddock look like? About 40 yards X 30 yards. The more square you make it, the more evenly the animals will eat and the less trampling and walking they'll do. They'll settle faster and be more content.

What if you only have 2 cow equivalents and the grass is really, really good, up there at 80 cow-days per acre?. Although that means one acre would feed 80 cows for one day, you don't have 80 cows. You only have two. So all you need is 1/40th of an acre. Follow that? Well, what does 1/40th of an acre look like? Let's see, 1/40th of 5,000 square yards is 125 square yards. What does that look like? If you said about 10 yards X 12 yards, you'd be right. I know my numbers aren't exact in these examples and I'm doing that on purpose to get you into a mindset that this is like horseshoes: close counts.

You don't have to be perfect, just close enough. Now, think about this. If you have an acre that you divide into 40 paddocks and that's all the land you have, how long is your rest period? If you said 39 days, you just qualified for graduate school. The rest period starts when you vacate the first paddock and ends when you re-enter it. Can you imagine one acre with 40 squares in various growth stages? It would look like a quilt, wouldn't it? That is exactly what your pastures will look like under this program.

They won't be grazed into the dirt. They won't look like a golf course or a lawn. They'll be a patchwork of varying maturities which offers varying amount of cover and habitat for spiders, moles, voles, grasshoppers and bird nests. Management intensive grazing is not only good for the land and the animals, it's also good for every kind of critter you can imagine. All those critters lend stability and abundance to your homestead.

One of the most common concerns from people starting this protocol is "how much do I give them?" If you've never allocated a day's grazing, how do you know how big an area to offer? First, you can just guess. You'll know a lot more in a day than you do right now.

Another starting point is to ask another farmer or the county extension agent what the normal acres-per-cow stocking rate is in your area. If it's 4 acres per year, divide that by about 250 days (this is exclusive of hay feeding) and multiply by 4 grazings per year to get a feel for what's likely. If the rainfall is less than 20 inches, multiply by only 2 grazings per year. If it's less than 10 inches, it'll only be one grazing per year. This may appeal to you engineers, but I can assure you, just giving a guess and making the adjustment tomorrow seems easier to me.

The next question is "how do I know they're happy?" Well, you ask them. Ha! Contented animals won't stalk the fence or bellow. They spend a lot of time lounging. They settle. Another great way to tell is to monitor their manure. If it's juicy, their feed is too rich. If it's hard, it's too fibrous. For cows, this is expressed as sheet cake or cookies. What you want is a pumpkin pie:

round, standing up about two inches, with a slight depression in the middle and a little standing drip where she pinched off the last bit. Yes, indeed, cow pies tell the story.

Look at their left side where the paunch is. How indented is it? Does it look curved and filled? With a little practice, you can definitely tell whether an animal is full or not.

With all that said, sometimes a homesteader will have only an acre but still want a milk cow. An acre won't support a milk cow. It'll come close, though, especially if the cow is of a small stature, like a Jersey or Dexter. In these situations, you may want to hybridize by strategic grazing and hay. Normally you don't want to feed both because the digestive microbes vary with the type of forage.

But if you run low on grass, rather than continuing to graze it before it can recuperate, you gain far more total grass to pull the animal off and feed hay for a couple of weeks or a month to let the grass recuperate. Remember, it's all about encouraging the fast growth curve. In a dry late summer, we sometimes feed hay for a couple of weeks until the fall rains come and the grass starts growing rapidly. By holding the herd off the grass completely, we never graze it too short and it recuperates much faster once rains come. By feeding hay for a couple weeks we might gain a couple months of grazing time.

One of the most strategic grazing practices you can utilize is taking off the pruning pressure during dormancy, whether that dormancy is caused by drought or temperature. You could confine the cow to a piece of brambles you're trying to stomp out; feeding hay there and concentrating her impaction can do a lot of good in a dry time. Or you could feed the hay in a shed. If it's hot and dry, your cow may enjoy being in the shade on clean, deep bedding. When rains come, you'll have compost to spread.

Having gone through this, I can appreciate feeling intimidated if this is your first introduction to an ecology-based grazing program. The ecology of it is overwhelming; no argument. But do the ecology benefits translate economically?

To be sure, the economics on this look better and better the larger the farm's scale. Every herd or flock needs to be surrounded by fence, needs access to a water trough and mineral box, and depending on season, ability to get to shade. Those requirements are the same regardless of size. Therein lies some of the pushback to be this meticulous or anal about management if you're a small producer. I can hear folks say, "I just put my two steers out on that three acres and they seem happy to me. Why all this fuss?"

My first response is to ask you about your stewardship ethic. Are you really interested in ecological authenticity or not? Do you really want a resilient environment around your homestead or not? One problem with all this is that nature is incredibly forgiving. We can punch it and punch it and it keeps trying to produce, keeps trying to take care of us. Even under incredible abuse it keeps trying to sustain us and give us what we don't deserve. Generally degradation doesn't happen overnight. The slowness takes the edge off the obvious.

But if you have a flock of goats out there on 5 acres year-round, or any other animal for that matter, I can guarantee you that the production is dropping every year. The diversity is dropping every year. The resilience is dropping every year. For no other reason than ecology, anyone who cares about stewardship should have a burning desire to manage with ecological integrity.

My second response is economic. In our region, the average cow-days per acre is 80; the average pasture acre in our county can feed 80 cows for one day a year or one cow for 80 days. Got that? On our farm, we average 400. On our partner farms that we manage for other owners, we average 240. Both of those numbers are more than double the county average. Here is why the production goes up.

If an animal grazes a blade of forage as soon as it's grazable, in its infant stage--remember the three stages?--in our area that represents in the neighborhood of 20 bites per year at about 200 pounds per acre, which is a total of 4,000 pounds of grass per acre. By limiting the bites to four and letting the juvenile growth stage catapult the biomass to 3,000 pounds per acre before grazing, that totals 12,000 pounds per acre. The oft-bitten grass never gets out of infancy; it never gets out of first gear. But by controlling the meeting between pruner and vegetation, we give that plant time to go through its juvenile growth spurt.

By making sure that the plant spends more time in its rapid biomass accumulation phase, we inherently enjoy more production. Let's say you went on a trip and you could never go faster than 10 miles per hour. But then you figured out how to go 25 miles per hour 60 percent of the time. Would that be a game changer? Absolutely. That's what giving the plant the opportunity to grow faster offers. During the times of juvenile growth that you're now giving it, the plant accelerates its sunbeam conversion rate into overdrive.

That's just the production angle. When you add the utilization rate due to more aggressive grazing, more uniform grazing, and less walking around, that efficiency adds yet more cow-days. How much time does it take? On our farm, our benchmark average is 30 minutes per move. That includes moving the herd, taking down the back fence and resetting it as a catch fence, moving the mineral box and moving the water trough.

At 30 minutes per day, it doesn't take much increased production to make the economics look good. The last time I ran a return on labor, it approached $100 an hour. I don't know about you, but I can work for that.

We plan ahead to minimize moving time. For example, if we're moving to adjacent paddocks (sometimes we move long distances to another field) we position our water trough under the cross fence so they can drink out of one end one day and out of the other end the next day. That means we don't have to move the water trough every day.

If you're pressed for time for a couple of days, you can either not move them every day and give them a two-day or three-day paddock, or you can set up the paddocks beforehand to clump your fence building and removal to one trip out, maybe on the weekend. Teresa and I received a conservation award when our son Daniel was only 8 years old; we had to be gone for two days to receive the award. I set up the fences and Daniel, at 8 years old, moved 100 cows each day while I was gone.

Moving animals to a new paddock is not strenuous work. You can continue doing this even after you're no longer able to drive a car. You can do this when you're barely out of diapers. Or when you go back into them. This is mental work. The trick is in knowing how much to give for a day and that comes with experience. So what would it be worth to you to be able to double or triple your production from pasture on your homestead? Could you afford to devote half an hour a day to make that happen? Do the math. I don't think very many folks would say it's not worth it.

The reason I'm adamant about daily moves is because that's the fastest way to learn. If you move only weekly, in a year you'll have 52 tests. If you move daily, you'll have 52 tests in 52 days. That's a vertical learning curve. Mastery comes fast when you're taking a daily test. In a couple of months you'll settle in and feel comfortable and one day you'll suddenly have the revelation that the lackadaisical neglectful convenience thing is for losers. You'll realize you're on the winning track and can't imagine not going out and moving the animals. That's when you know you're on your way to success.

The final benefit of this constant movement is harder to quantify, but it's no less valuable. When you see the animals every day, watch them pass by you on their way to the new paddock, see how they walk, who holds back, their poop, their eyes, their hide coat, you see things you'd never notice if "they're just out there doing their thing. I guess they're fine." One of the first big information investments I made was to attend a cattle school in Kansas. The instructor said "I could tie a milk jug to the tail of all your cows and most of you guys, the way you check your cows, you'd never even notice."

He was exactly right. When you're among them, moving them, watching them respond to you, your observations are far more acute. In addition, your animals get used to you. They don't run away; they come to you. They associate you with good things--a new salad bar. On our farm, this creates a kind of game, to see how close we can get to the herd before they see us and respond. Remember, animals love routine. Try to move them about the same time every day. Don't throw them curve balls.

And when it's time to get them into a corral, they'll follow you in because they're used to trusting you. I can call my cows and they'll follow me into the corral with nobody pushing from behind. This daily interaction builds trust and that makes for happy animals. When your animals feel secure in your management and care, they'll be happy. They know you'll see when things aren't right. They know you'll notice a limp or a cloudy eye.

All of that trust and relational equity translates to better stockmanship, which always adds to the bottom line. I'll end this chapter with a warning: this is not a fencing scheme. Yes, it involves fencing, and we're going to drill down on that. But management intensive grazing uses fencing as a tool, as a control mechanism, for all the beautifully beneficial reasons I've described here. This is a management protocol that is primarily mental, mathematical, observational, and creative. Fence is merely a tool that enables us to leverage all those attributes.

Chapter 13

Fencing

Controlling your livestock is more important than genetics, forage species, or equipment. If you can't keep them home and put them where you want them, you've got nothing but headaches. The two types of fencing are permanent and temporary (mobile or portable). The two basic types of permanent fencing are physical and non-physical. A physical fence is a structure that an animal can't penetrate. A cow can't knock it over; a goat can't go through it. The animal can push, shove, dig, jump, but can't go through it, under it, or over it. That's a physical fence.

Non-physical means the deterrence is psychological. That is normally some sort of electric shock. Whether it's buried dog fences or 10,000 volt multi-strand electric fences to control a rhinoceros, these fences rely on pain and memory in order to work.

Both have their place on your homestead and both come in many different shapes and sizes. First, let's discuss physical fences.

I like these for boundaries because they don't depend on electricity to keep them going. While it's true that a tree can fall on them and smash them to the ground, they are not subject to conduction problems. If they are up, they work. If they're down, the breach is easy to see.

By far and away the most expensive boundary fence is a board fence or any type of wooden fencing. While these are pretty when first installed, they rot, warp, and require substantial maintenance to keep up. Perhaps their only saving grace is that they do not require brace posts since nothing pulls on the corners. Further, a board fence will never keep in chickens, pigs, or lambs.

Different animals require different types of physical fences. Big animals like cows need something strong but it can have fairly large web squares. Small animals like chickens need something almost impermeable but it doesn't have to be as strong. Physical fences that can control all animals are usually some sort of chain link design and of course extremely expensive. A board fence will never keep a chicken in and a poultry netting fence will never keep a cow in.

For goats the saying goes something like this: "if it won't hold water, it won't hold a goat." Remember that animals don't have dental appointments or go to school or church meetings. They have 24/7/365 to explore, poke, dig, scratch, rub and otherwise test a fence. While you're sleeping or doing other things, they're out there, probing and cogitating. When an animal looks at you with those big penetrating eyes, don't be fooled. It's thinking "the grass is always greener on the other side of the fence."

In a bit, we'll discuss all the things necessary for an electric fence to work. Suffice it to say I don't want spark contingency on my boundary fences. You can't have stress-free nights of sleep unless you know your animals are contained and staying home. To be sure, everyone who has had animals has had an errant situation, but these should be rare enough that they occupy only the long-ago memory "remember when . . . ".

The old adage "good fences make good neighbors" is still true. In general, fencing law requires you to keep animals home. Two types of laws apply to livestock: inclusion and exclusion.

Back in colonial days when fencing was practically non-existent, most jurisdictions adopted exclusion laws. In other words, if I didn't want livestock trespassing on an area, I had to fence them out.

Animals could kind of go wherever they wanted, even across property lines. If I didn't want them on my property, I had to fence them out. Early fences, including those made of stone, were primarily exclusions to keep animals out of gardens, orchards, and crop fields. As the country became more populous and wire cheapened fencing costs, more jurisdictions adopted inclusion laws.

Inclusion meant that you had to keep your animals on your property. In other words, if my animals gets on the neighbor's property, it's my fault regardless of anything, unless they went through a legal fence. Most states have requirements for what constitutes a legal fence, but in convoluted legalese it's circular. The definition for a legal fence is one that contains livestock. If it doesn't contain livestock, then it's not a legal fence.

If my animals escape from a legal fence, I'm not liable for whatever damages they cause. But if they go through the fence, it's not a legal fence. I've always thought this legal mumbo-jumbo comical. The safest way forward is to assume that you need to keep your animals home and if you don't, you're liable. A permutation of this theme is on boundary fences where one neighbor has livestock and the other one doesn't.

In most areas, if your neighbor doesn't have livestock, you have to build a fence to keep yours in. The neighbor has no obligation to share costs of the fence if he doesn't have livestock. However, if he has no livestock and you build a fence because you want livestock, and then he decides to get livestock, he's obligated to pay for half the fence. If two neighbors have livestock and need to build a new or replacement fence, the laws are fairly specific that each must share half the cost. While this may sound simple, it can still get complex if the two neighbors can't agree on what kind of fence to build. Size of posts and type of wire can both have a big impact on cost.

Another complexity issue is what happens between builds. One of the biggest costs of fencing is not the initial build, but the ongoing maintenance. If you and the neighbor agree to build a fence and split the cost 50/50, what happens two years down the road when a windstorm blows a tree over and mashes it down? Then what?

Because of this conundrum, most rural areas have a kind of neighborly agreement to split boundary fences in half. In our area, the most common is to find the midpoint and each land owner takes the half on his right. In other words, if I share 1,000 feet of fence with a neighbor who has livestock, we measure 500 feet from the corner, each stand on our side of the fence, and take the 500 feet from the midpoint to our respective corner. That way we each build and maintain half of the fence.

This works fairly well except for cases when one half receives heavy wear and the other half doesn't. For example, we share a fence that bisects both open land and woods. Both of us have cows. His cows have free range in both his field and his woods. On our side, we have an electric fence that excludes our cows from the woods. The open land, then, has cows on both sides but the woods only has cows on one side. The wear and tear on the woods is not nearly as much as it is out on the open land where his cows and our cows try to fellowship through the fence.

In this case of mid-point division, our half was through the woods and his half was through the field. Over the years, we've done maintenance on his half because it gets much more wear with animals on both sides. This is just courtesy. I don't know that anything formal exists to compensate for this kind of disparity. If we wanted to be persnickety in this instance, we could tell him we will never have cows in our woods and he would be obligated to keep the boundary fence up to keep his cows out of our woods.

That position, though legal, would be perceived as ill will. If our cows ever breached our electric fence, which has happened, and got into our woods and either got over on him or he saw them in our woods, he'd have a legitimate reason to think ill of us for not sharing the cost of that woods fence. It all kind of evens out. Fence building is easier on open land than through the woods so the initial build is problematic, but if cows are only on one side it adds many years of life to the fence.

We have another neighbor who reversed it, preferring to take what was on the left instead of what was on the right because his left half went through an open field and the other along our woods. He preferred to take the extra wear and tear of animals on both sides rather than the difficulty of putting in posts where tree roots would make digging difficult. He also didn't want the chore of fixing it when trees or branches fell on it.

And then we had another neighbor (now deceased) who wouldn't ever fix a fence for love nor money and we ended up building his half and our half just to keep our animals home. That deteriorated into a bad situation, which brings me to the point that perhaps no issue creates more rural tension than animal trespass and fences. I've gone on at this to some length because I don't want you wrangling in court or coming into a community on your new homestead unaware of common courtesy. Always err on the side of doing extra and you'll have happy neighbors and respect in the community.

The main thing to remember is that you must keep your animals under control. At the end of the day, if you have a dilatory neighbor who doesn't want to share, you still have to keep your cows home. To my knowledge, no legal recourse exists to force someone to pay their half of the boundary fence. That said, if the neighbor's animals come through his poor fence onto your property, you can impound them and demand a ransom payment. Unfortunately, we've had to do that a couple of times through the years with habitual offenders. It does indeed get results.

When you start controlled grazing, often you'll have grass when the neighbor doesn't and that puts more pressure on the fence. His cows look over and see your nice grass and start pushing on the fence to break in. One of the most dramatic stories I ever heard was a guy whose neighbor's bull kept coming over. He finally got the bull into his corral and into his head gate and pulled down hard enough on the bull's testicles to break the chords that send sperm up to the penis. He took the bull back to the neighbor who always wondered why none of his cows got bred that year. That's ornery, but it does illustrate the level of frustration that can build over fence issues.

Few things bring as much satisfaction as a well-made, well-maintained fence. Few things bring as much good will among your neighbors as a well-made, well-maintained fence. Commit today to not be the negligent ne'er-do-well who is constantly gathering animals from the road and the neighbors. When you get sick and need some help, or when you need advice, you'll have a hard time of it if your neighbors have gotten sick of you due to poor animal control.

Besides rubbing from animals, the other primary reason fences deteriorate is vegetation. Birds love to sit on fences. Birds poop where they sit. Because birds eat a lot of seeds, their poop contains a lot of seeds. Are you following where I'm going with this thread? Fences attract every bush, shrub, thorn, tree, or weed that finds a home in the area. The more obnoxious the plant, the closer to the fence it wants to be. You'd think a sapling wouldn't want to grow up through the webs of a woven wire fence, but for some reason, a young tree much prefers that to growing up a foot away where it can become a beautiful, healthy tree without wire constraining its growth.

Because trees grow only at their tips, their trunks expand horizontally, not vertically. If a tree grew up all over, it would gradually pull the fence up. But no, it swallows the fence at height. That's why you see pictures of trees growing up through the hoods of cars in old junk yards. Trees eating wheels and old machinery as they expand in girth are quite amazing.

Because of that biological phenomenon, many folks are tempted to use trees as surrogate fence posts, nailing wire directly to them. I'm having a hard time controlling my fingers as I type this sentence. I will make this bold: FEW SINS ARE AS GREAT AS PUTTING HARDWARE IN A TREE. I hope that was bold enough. Never, ever, ever, ever, ever hammer a staple into a tree. Further, do all you can to make sure no wire ever gets in a tree.

First, hardware injures a tree. It has to grow around the hardware, then gradually absorb it. The scar is permanent. A wire through the middle of a 24 inch diameter tree will wick black iron oxide a couple of feet up the trunk, destroying the quality of the lumber if and when the tree is harvested. The second reason, though, is about chainsaws and personal safety. If you've ever hit a piece of wire in a tree with your chainsaw, I don't need to explain what it does to a chain. Further, if hit just right, it can break a chain, send it flying into your leg, and you to the hospital emergency room.

Radical environmentalists spike trees on logging projects they oppose because they know the thousands of dollars and personal injury that can occur from putting hardware in trees. Remember, trees gradually eat what's on the outside today. In a few years, it'll be inside as the girth widens, imbedded in the tree sometimes only discoverable by a chainsaw half a century hence. Around our farm, the unforgivable sin is putting a nail, staple, wire--any hardware--in a tree. Do the right thing by the tree and your grandchildren--set a post.

This also means trying to keep trees away from your fence. I like enough room to drive on both sides of a fence. That way I can check it but it also means I won't ever have trees with imbedded hardware. If I'm building a fence along woods, I always clean back about 20 feet in order to give me at least a few years not to have to worry about a tree or branch falling on my new fence.

Physical fences are expensive to build; you want to buy as much maintenance-free time as possible. Sometimes that means negotiating with a neighbor to get permission to harvest some trees. Normally, reasonable people are thrilled enough with the prospects of a new fence, and appreciative enough of the effort, that they'll grant permission to take down anything near the fence line. Only a couple of times in my experience have people refused to let me take fence line trees down as part of a new build. If you're building a fence along a woods edge, the trees always reach out to the daylight. They often grow extremely well, too, because of the extra light afforded by the field edge. You never want to build a new fence with a bunch of massive trees overhanging it.

With that said, most boundary fences are some type of woven wire or high tensile or high tensile webbing. Lots of different types and styles of fence exist. Among the various webbed wires I don't have a huge preference. I don't like the multi-strand high tensile fence (the cheapest kind) because although it's touted as physical, it's really not. In New Zealand, where it originated, it always has a few strands electrified.

When that kind of high tensile option came to the U.S., farmers' aversion to electric fences meant that these fences were installed as physical fences without any spark. As cows stick their heads through the strands, pushing and shoving, they loosen the wires and break the timber stays between the posts. The whole idea is to reduce posts and use cheaper wooden stays (often called Insultimbers) in between. But they crack and break and then suddenly the span between vertical positioning extends all the way to the fence posts.

The only time I saw this kind of high tensile fence work was when a neighbor used tight 8 foot spacing between his posts. He put it along a road because the snow plow constantly pushes snow into the fence. Having no vertical wire, the horizontal-only wire lets the snow flow through better and reduces his maintenance (fewer bent-over posts). But he didn't depend on those little timbers between posts; he used tight post spacing and he's also an excellent fence builder. In a similar situation, I'd use five strands of barbed wire rather than nine strings of high tensile smooth wire. He doesn't have sheep so spacing can be fairly wide.

The key to a physical wire fence holding up is the corner bracing. The main thing to remember is that you can never have too big a corner post. Get the size you think you'll need, then add a couple of inches. And put the freakin' thing in the ground. I mean bury it. Then put in a horizontal brace (post connecting the corner post to the second post) that will stay there, not one the cows will rub out. The brace wire needs to go to the BOTTOM of the corner post. The idea is that as you put tension on the fence and the corner post wants to pull over, it can't because the horizontal brace post and the diagonal brace wire won't let the post slant. It can only pull through the ground vertically, which of course it can't do.

The Spanish windlass is the way to make sure your brace wire is tight. Or you can use those handy dandy tighteners. I'm old school and still prefer the windlass--it's cheaper. The key is to make your brace wire as tight as possible so it doesn't have much slack. That way when you put your stick in between and start to turn, it'll tighten up right away. If you have too many turns before it gets tight, you'll break the brace wire. I only use Number 9 wire for braces; it's tough and a bit hard to work with, but it'll stay there until the cows come home.

Don't skimp on your corner or second posts. Make them beefy and lined up with the horizontal braces. Fasten the horizontal brace well. My favorite technique is to use a chainsaw to carve out a notch in the corner and second posts about chest high. I place the brace snug into those two opposing cut-outs and then using a drill, make a toenail hole in the brace. This keeps the end from cracking when I pound a spike into the brace posts. I put one on both sides of the brace to really hold it in place. Don't skimp on your bracing. It's the key to everything. Why go to all that fence building effort only to short-cut the anchors? Do it right and you'll have a fence that lasts for a long time.

A physical fence has two parts: the wire and the posts. Let's put some attention on posts. Obviously you want something that won't rot. I can't tell you how many newbies I've visited who, in trying to save some money, used whatever trees they could find for posts. Big mistake. To my knowledge, in North America we only have three basic varieties of trees that are suitable for posts, unless you buy pressure treated posts from the lumber yard.

Here they are: locust, osage orange, and cedar or fir. Obviously redwood can work. We have locust, osage orange, and cedar in our area, but even among them, some are better than others. With cedar, what you want is something that's primarily red. If it has a lot of white before getting to heartwood red color, it won't last but a few years. If it's mostly dark, deep red colored inside, it'll last many decades. I know some cedar posts on Teresa's grandfather's farm have been there nearly 100 years and they're still sound.

Osage orange is great but it's hard to find a straight piece. As far as I know, all osage orange is fine. The different qualities like we see in cedar and locust don't seem to exist in osage orange. Every tree is usable. But it's hard to find a straight piece long enough for a post. And with its stickers, it's a hard tree to work with, but it's a tough wood.

In our area, locust is the old standard for on-farm posts, and it's certainly the one I prefer. But they are a little like the cedars in that two trees growing a few feet apart can exhibit profound differences. My tree identification books do not list a yellow locust; they only list a black locust. But any old-timer (now I ARE one) will tell you that only yellow locusts are worth planting as fence posts. I've certainly found that to be true.

When you cut a locust, if the inside is yellow, it's a keeper. If it's black, cut it up for firewood. Again, like the cedar, I know of locust posts that are many decades old and still sound. A good yellow locust lasts way longer than anything pressure treated. A black locust is still rot resistant, but it'll only last about 10 years before deteriorating. Again, folklore has it that other critical factors include cutting the trees with the sap down (in the winter) and in the down side of the moon (waning moon rather than waxing moon). This may sound a bit superstitious, but I admit that I try to adhere to both of these principles when cutting post material, especially the sap down idea.

Now a couple of installation tips. Bark, regardless of tree type, does not last long. In climates with more than 20 inches of rainfall, posts rot at ground level; that's where the biological activity is. In brittle environments with low rainfall, rot develops at the end; the post actually oxydizes rather than rotting. Rot cannot happen when oxygen is excluded. If you don't take the bark off the post, it forms an oxygen conduit that stimulates rot. Debarking is important.

You can sharpen the end of a post with a chainsaw and pound it into the ground with a post pounder. Obviously the time to install posts is when the ground is damp and soft, not when it's dry and hard. I've dug a lot of fence post holes by hand and if you're not hitting a bunch of rocks or roots, it goes along well. I've certainly put in four or five posts an hour by hand when the digging is nice. In any case, keep sod out of the hole--again, this creates air pockets that encourage decomposition.

I always install posts with the small end down. Look the post over carefully to determine which way the branches were growing, then install it upside down to its natural growth. The reason is to shed water, like a shingle, rather than let the water stand in those crotches and gradually create rot spots. Anywhere water stands or air penetrates will speed up post decay.

One of the best things you can do to make a post last a long time is to cut the top off when all is done. After the fence is pulled, I go along with the chainsaw and cut the post tops off (it also makes the fence look a lot prettier) facing the south, on a diagonal. The diagonal cut makes the water run off and the southern face makes it dry out faster after a rain. Moisture also makes things rot. Preferably, the side toward which you drain the water will not be the side on which the fence is nailed, but sometimes you can't have everything in your favor.

I never put a fence hard to the ground because I want enough space between the bottom and the ground to get the snout of a chainsaw in and cut saplings. Remember that over time the soil will *uppen* under a fence due to the extra vegetation attracted there. Because you can't mow right up to the fence and the fence itself acts as an energy antenna, over time soil builds up under a fence. Leaving a few inches of extra space will serve your children well when they continue to maintain the fence.

Many times the biggest part of a fence building project is actually the preparation work. Getting the site opened up and cleared, pulling out the old rusted fence, removing trees with imbedded wire in them--all this can dwarf the actual building project. Again, please don't just leave the old fence and build a new one four feet away. That's a blight on the landscape. Go to the effort of ripping out the old fence, even as entwined and junky as it is, to create a nice clean space for your new fence. Don't create a problem for your children. Do it right. Over the years, much of our firewood and many of the logs we mill on our band sawmill come from our fence building projects. Few things are as gratifying as cleaning up an old broken down fence line and replacing it with a new, tight fence.

If you're putting up any of the webbed wires, which is certainly my preference, don't make too long a pull. You can pull a fence tight enough to break it. Normally these rolls of fence come in about 320 foot lengths and that's about the right length for a pull. You can fudge to about 500 feet, but if you go over that, do an intermediary pull. You don't need to install a whole brace assembly on the intermediary pull.

You can create a poor-boy simple brace by notching the temporary end post and putting a post on a diagonal down to the ground. This will hold the temporary pull post long enough to scramble to the corner post and do your final pull. Obviously if you do need a two-stage pull, make sure you pound your staples in hard enough so that when you let up on your pull, the wire can't slip through. We usually put in two staples just for security. Always wrap your corner posts with the wire. This takes time, but it's the only way to be sure it's fastened permanently.

If all you're running is cows, you can fudge on the physical boundary fence. I wouldn't do this with anything else, but because cows require the least electric fence, occasionally we buy boundary fence replacement time by offsetting an electric. Normally an old overgrown fence is a bit of a barrier anyway due to brush and vines. Putting an offset electric three or four feet away can buy a year or two until you can get that boundary fence replaced. We do this routinely on rental farms.

On our farm we also run pigs without physical boundary fences, but these electric fences must be multi-wire and highly maintained. Just like in the discussion on physical fences, I'm not going to tell you what brand to buy. I'll be far more basic than that. Here we go. I've listened to countless farmers and homesteaders lament to me "electric fence just doesn't work for me." I have a secret message for all these folks: "you're fudging. It works; it's up to you to make it work."

Because an electric fence is a psychological barrier, it depends on several requirements.

1. Visibility. If an animal can't see it, the fence doesn't exist. Not only does this mean the wire must be bright, it also means it can't be buried in vegetation. This is the bane of electric fencing. A physical fence, even when covered in vegetation, still works, unless the brush and vines have pulled it down. But as long as it's upright, it'll stop things.

An electric fence, on the other hand, is useless if the animal can't see it. Most animals have good eyesight; pigs have the worst. That's problematic because pigs need the lowest wire and they're constantly stirring up dust. Pig wires are the hardest to keep visible, but keeping them visible is absolutely essential if you expect the pigs to stay in.

This is the major reason I don't like any steel electric fence wire, including high tensile. Anything galvanized, regardless of coatings, will eventually turn dull. When the shine wears off, you've got a gray camouflaged wire that's hard to see, especially for wildlife. The reason deer run through electric fences is not because they have a vendetta against you; it's because they can't see them. That's not always the case--sometimes they run away from something in a panic and fail to see even the most visible fence. But generally, if they can see it, they avoid it.

For permanent electric fence, I like high tensile aluminum wire. We prefer 12.5 gauge. Aluminum costs more than steel per linear foot, but it weighs about a third as much and conducts 30 percent better. And it never rusts. For the dubious, I invite you to our farm after dark; walk out through the fields and you'll see 30-year-old aluminum wire glistening in the moonlight. Although the moon doesn't shine every night, it shines enough for the local wildlife to remember where things are. The deer become familiar with every path, every tree, every puddle and they certainly know where those fences are.

Premier Intelli-rope works great as an offset when we're trying to get onto a place with broken down fences and don't have time to repair them all right away. It's a quarter-inch highly visible, strong poly-wire that has never failed us unless somebody accidentally shoots it (that's happened twice). The more risky your situation, the more visible you want the electric wire. That's why you routinely see wide poly-tape around horse paddocks; not because the horses don't respect electric fence, but because they need to see it dramatically.

While a physical fence can be well maintained with a once-every-five-years schedule, an electric fence requires almost annual maintenance. In addition to compromising visibility, vegetation also reduces the spark voltage with little shorts, or faults. If you take a walk after a rain, sometimes you can hear the faint pop-pop of a spark hitting a stem or leaf. That drains away the voltage and keeps the current too low to be effective. Non-physical fences are not self-maintaining. Although they are cheap to install, they require constant attention to keep them clean.

2. Tightness. All animals carry a bit of insulation in hair, wool, fur, and feathers. If a wire has sag, not only is it not at the proper height, but more importantly, it can give way when an animal inadvertently rubs up against it.

The harder an animal has to push to get good contact, the less effective the wire will be. Think about pushing against a loose wire versus pushing against a taut wire. One might give several inches before becoming tight enough to get to the skin of an animal, where conductivity occurs. A taught wire, on the other hand, makes immediate contact. This keeps the animal from playing around with it or pushing the system.

To be sure, the wire doesn't have to be as tight as a guitar string, but it does need to be fairly tight. All of the polywires and polybraids for temporary wire have a lot of stretch. I virtually never use those for permanent fencing; I prefer the aluminum high tensile. You can pull the 12.5 gauge wire as hard as you want without breaking it; ideally, on a cold winter day you want to hear the wires sing when a bit of breeze goes through them.

With a single strand and some beefier corner posts, we can get this degree of tightness without any bracing using aluminum wire. With steel wire, even with one strand, you'll need a brace to keep it tight. I've often watched folks try to compensate for loose wires with extra posts. That never works. You don't want that wire to move at all before the animal gets a shock; if it has to move, the animal won't fear it as much and can slip over or under on its way to contact. Tightness has a lot to do with sag, which has a lot to do with height.

3. Height. The vulnerable part of the animal is its nose; that's where the least insulation protection is. It's also the point of inquisitiveness and curiosity. When unstressed, animals approach an electric fence cautiously, just like they do anything that's new in their world. They don't go busting through it if they're calm.

Think about the animal(s) you're trying to control and where the nose is. Measure some if you can't figure it out. The smaller the animal, the less the tolerance. A big cow, well trained, will respect an electric fence within 18 inches of tolerance. A lamb, on the other hand, or little piggy, might only have a couple of inches of tolerance. That's why we use two strands for pigs up to about 150 pounds and then shift to one strand.

In addition, different animals have different tendencies for going through a wire. All things being equal, a pig would rather go under a wire. A cow would rather go over a wire. Sheep and goats tend to go over as well. In other words, if you're going to err, err on the side of low for pigs, but err high for sheep, goats, and cows. Remember that as animals grow, not only do their tolerances increase but their discipline to the wire increases as well. A rambunctious young animal will test a wire routinely; an old one won't.

Well disciplined older animals are much easier to control than adventurous and mischievous young ones. That means that you can push things a bit with older animals. For example, if you have the chance to expand your operation onto a neighbor's vacant field surrounded by marginal physical boundary fence, you might consider wrapping it with a single strand of electric fence and using it if you have a well-disciplined group of cows. But if you've just bought some stockers, you might not want to use that field until you have a physical fence barrier around it.

4. Voltage. Unlike humans with shoes, animals are always well grounded. The way electric fence works is that the energizer sends a pulse out through the wire that doesn't complete a circuit until something touches both the wire and the ground. At that point, the electrons complete the circuit to the ground rod and that's when the critter (or you) feel the shock. That's why a bird can land happily on a 50,000-volt high tension power line. It's not grounded.

In order for electric fence to be effective, it needs to be hot enough to get the animals' attention. That varies from species to species and depends on how well trained an animal is. Well trained animals don't need as hot a spark as untrained. When training animals to electric fence, you want their first experience to be memorable.

For pigs, we put a wire across a corner of their receiving pen with a heavy spring in it so if they hit it they don't break it. With only a few feet on a 1 joule energizer, we can get 10,000 volts on that little stretch of wire. It'll throw a spark a quarter inch but that's what you want in order to keep them from ever challenging that fence again. Their respect for the wire is what enables them to enjoy the pasture for the rest of their life.

For cattle, we do the same thing; a fence across a corral corner. You need to place it out far enough that it's not just a couple of feet from the physical fence; you need to bring it out far enough (at least 10 feet) so they have plenty of room between it and the physical fence. They need license and incentive to want to go through it; if it's too close to the physical fence, they have no desire to get to the other side of it. These tightly controlled procedures protect you from the animals getting completely away even if they go through the training fence.

In general, I like anything above 3,000 volts as a maintenance spark. The most important tool you can buy for electric fencing is a good fault finder. It will tell you the voltage and how big your fault bleed is and what direction it is. That's invaluable because with electric fence, you're often chasing down things you can't see. You can't see electricity as opposed to a water leak.

5. Consistency. Even the best trained animals in the world will occasionally check an electric fence. The most important trait of good fence is that it carries spark 24/7/365. If you get careless and quit checking it with your fault finder, or let the energizer quit functioning, inevitably your animals will begin testing the fence. If you don't, they will.

You be the one testing it, every day. No cheating. Don't ever walk away from it not working, cavalierly telling yourself "oh, it's okay, they'll stay in." Big mistake. If you're using a battery and it's blinking red, change it out. If the fault finder says you have only 1,000 volts, go find the problem. You'll be sorry if you don't.

Those are the critical elements of permanent electric fence function. Now for some catch-up rules:

1. Don't be cheap when buying an energizer. I tell folks using electric fence that the one place to invest is in the energizer. Cheap ones don't last and don't work. Get the good ones.
2. Get energizers from professional suppliers, not the local farm store. I wish I didn't have to say this, but unfortunately your neighborhood farm store's electric fence supplies are not geared for serious practitioners.
3. All that matters is joules. If your energizer says "will charge up to 10 miles" you know you've got a piece of junk. No reputable energizer manufacturer puts those

sorts of claims on an energizer because people who know anything about energizers know that the power is determined by joules. Joules measure the energy produced.

4. Make sure your ground rod is adequate. The more fence you have out, the more grounding you need. Better to err toward too much grounding than not enough.
5. Consider complete mobility rather than central systems. When we began leasing properties, we converted even our home system to 100 percent mobile energizers and we love it. We use simple 3 ft. ground rods, solar energizers from Premier, or marine battery-powered energizers, and move them with the animals. Goodness, we have 15-20 in operation at all times, but we don't have to keep up a massive grid. It's easier to run down faults and it's certainly nice to know that visitors' children won't get shocked if they touch the wires most places. We just tell folks "the wire is hot next to animals; if you don't see animals, the wire is probably okay." That puts parents at ease.
6. Stay away from metal posts with permanent electric fence. Don't use T-posts. We use 3/8 inch rebar for our mobile stakes, but permanent fencing is not something you see or check everyday. If an insulator fails and the wire touches the metal post, you have a dead short. Always use some sort of wood so even with catastrophic failure, the fence won't dead short.
7. Never use barbed wire for electric fence. The barbs are for physical deterrence; they don't add anything to the effectiveness of electric fence. Keep your life simple and your fingers tear-free by using smooth wire for all electric fencing.

One final point on permanent electric fence. Sometimes you need to go through a ditch or a creek. In such cases, you probably don't want to put a post in the middle of the ditch or creek. We use a trapeze in these cases. We put a post in on either side of the low place and run the wire straight across. Then we put a second wire from post to post that sags down low enough to conform to the depression, using an old piece of metal or even a rock to hold it down. It just swings there like a trapeze.

In the case of a creek, we normally use two weights to make the wire conform to the banks and the water. When it floods, the trapeze wire either swings or a log takes it out. It always breaks away from only one side and swings with the current against a bank. In a few minutes after the flood goes down we can pull it back and re-install it. This works great with a second person on the other side of the creek. We use a piece of baler twine with a weight tied on one end to throw across the creek. Tied to the broken end of the electric fence, the twine enables us to pull the wire back across the creek and re-attach the broken end. Easy peasy.

Permanent fences consist of both physical and non-physical models. I like physical for boundaries but non-physical for all internal (except corrals). Well trained animals respect both.

Now let's talk a bit about mobile fencing. Different animals require different types of fencing. Here are my recommendations:

- Cows--one strand, nose height.
- Sheep--three strands or sheep netting (large holes).
- Goats--three strands or sheep netting (large holes).
- Pigs--two strands up to 150 pounds; one strand thereafter
- Turkeys: electrified poultry netting.
- Laying chickens: electrified poultry netting.
- Ducks: electrified poultry netting.

The reason for the large-holed sheep and goat electric nets is because the spark voltage stays higher. The smaller-holed poultry netting has miles of wire filaments that bleed off voltage through resistance. Fortunately, chickens respond to a lower spark. Sheep and goats require a higher spark; therefore giving up hole size for higher spark is definitely a good decision. The larger holed material costs less per linear foot too, so you save on the front end. We've used the smaller-holed poultry net for sheep just to eliminate the hassle of two different kinds of netting, but the sheep definitely respect the larger-holed and therefore higher-voltage netting better.

Turkeys are a bit different. Companies that make and sell electric netting generally make no guarantees about turkeys. Some even say "will not work with turkeys." We've been raising turkeys in electric netting on our farm for decades and here is what we've found. The companies are right: the turkeys don't get it.

For turkeys, the netting is a physical barrier; apparently they don't have enough psychology for a psychological barrier to work. We've been successful, though, at getting them to respect the netting by exposing them to it when it works like a physical barrier. That's when the turkeys are small.

We start turkeys in the brooder and put them out into the broiler shelters at about 3 weeks old. At about 6 weeks, we surround the broiler shelters with the electrified poultry netting, prop up the shelters, and let the turkeys come out. After a couple of days, we remove the shelters, replace them with the Gobbledygo (mobile turkey shelter) and we're up and running. The tricky part here is that if you put the turkeys into the netting too early, they go through the netting holes. We've pushed this by putting up two nets. The holes never line up precisely so this in effect cuts all the holes down to half-size.

At this size, though, the little turkeys are too susceptible to aerial predation. Giving them the protection of the broiler shelters gets them up big enough to be almost immune to hawks. By 6 weeks the turkeys are big enough to not go through the netting holes, but not big enough to walk over the net. At that point, the net is a physical barrier, not a psychological barrier. Fortunately, their inability to escape at 6-7 weeks stays with them for life and they just assume the netting is the end of their world, even when they weigh 25 pounds and could easily walk over it.

If you wait until 10 weeks or so to expose the turkeys to the netting, they're too big and simply walk over it. Amazingly, you can stand there and watch them get shocked. And even our turkeys, at 16 weeks, routinely reach through the netting to nab a grasshopper on the outside, touching their necks on the netting. The turkeys jerk with each spark pulse, but seem oblivious to it. They grab the grasshoppers and pull their heads back through the netting, jerking with each shock, but happy to enjoy the tasty grasshopper. All in a day's work.

The point of all this is that the turkeys have a sweet spot of exposure that makes them believe they can't get through the netting for the rest of their life. The spark is no deterrent, except for ground predators like coyotes, foxes, and dogs from outside. The spark still needs to be on to protect them from predators. But the spark has nothing to do with the turkeys staying in.

Chickens do indeed learn about the shock and stay away. Why don't they fly over? A lot of poultry psychology has gone into developing these electrified nets, not the least of which is to make them seem blurry. If you'll notice, all the brands use a varicolored scheme on the polyethylene webbing. The reason is that it's hard for the chicken to focus on this multi-colored webbing. For the chicken, it must be similar to a lion trying to focus on one zebra when a herd is running by. The blur makes it hard to focus and the lion can't distinguish one animal from another.

If you put a board at the top of the netting, all the chickens would be out in the morning because they would have something to focus on. Of course, the least whiff of a breeze sways the netting, leveraging the constructed varicolored pattern. To be sure, setting up this electrified netting is an art. Our procedure starts with laying it all out on the ground first. Then we start at one end putting in the posts with a slight outward slant. The nets are in circles and ovals; squares don't work well. The bend gives you room to stretch to make sure the fence is tight.

You can put together as many segments as you want to make as big a circle as you want. Just remember that frequent moves keep the animals content and clean; leaving them more than a week in one spot will make them begin testing the fence. New spots are always enjoyable for the animals and are all part of keeping them from pushing your fencing. Bored, discontented animals are always harder to contain than happy, satisfied ones. Frequent moves and new ground make a huge difference in the animals' testiness toward the fence.

One final note about fencing and control. The higher your animal density, the more pressure they exert on the fence. Electric fence psychology only works to a point. If you try to crowd them too much, or if you try to separate animals that really want to be together, their need to cross will eventually exceed their fear of the fence. The same is true with physical fences. A corral needs to be much stronger, for example, than run-of-the-mill boundary fences along a pasture.

That means that what might have worked fine for a previous manager or owner who did not practice high density management and movement might not work for you. Routinely, when our Polyface team takes on managing another farm property, we need to beef up boundary fences. Just because a conventional farmer neighbor says "the fences are fine" does not mean they are. They may be fine for a herd of 5 cows on 20 acres, but they won't be fine for a herd of 10 cows on a tenth of an acre. That's a huge fencing pressure change and you'll want to prepare accordingly.

Fences are one of the most important tools in your animal happiness tool box. When your control mechanisms fail and you get frustrated, the animals sense your frustration and then they get edgy. Enjoy building and developing fences that work in order to ensure happiness for both you and your animals.

Chapter 14

Grazing Development

When you're putting a book like this together, sometimes you have some things that don't fit neatly into one of the main chapters. This is where I'm going to pick up some loose ends. We've talked about basic design, the carbon economy, movement and fencing. That's all fundamental to effective animal husbandry. Now we'll touch on some more sophisticated aspects.

The first one is shade. We talked earlier about why it's important, but now we're going to look into the actual structures. Again, for the definitive information on this, please see the book *Polyface Designs*. I won't repeat all that extensive information here, but I am going to examine some broad principles.

Shelter has two extremes: hot and cold. But each animal has different tolerances and desires. For example, on a hot summer day nothing is more enjoyable to a pig or duck than getting drenched in a downpour. These animals relish the rain. Cows don't enjoy it as much, but they don't mind it. Turkeys don't like it but can tolerate it.

Chickens, on the other hand, absolutely do not like water. "Madder than a wet hen" is true. This difference in summer water tolerance affects the choice of shade covering. Everything but a chicken can get along well with a permeable covering, like shade cloth. Truly, shade cloth is one of the critical inventions that enabled pasture comfort and control to go hand in hand. Until we had shade cloth, we couldn't provide strategic comfort and focus both manure and impaction where it would all do the most good.

Stationary shade like trees has lots of limitations. The only farm animal not content with shade cloth is the chicken. Everything else is fine. That means that chicken shelter, both summer and winter, starts with an impervious roof. The difference between impervious and permeable roof is profound. A permeable shade cloth won't catch wind; a roof that sheds water won't let air go through it. That means you need structural design that can handle wind.

The critical design question on mobile shade is to make it light enough to move easily but heavy enough to handle wind. The best way to handle wind is with a low roof line. The lower the center of gravity, the better. The other consideration is the angle of the roof. If it's parallel to the ground, the wind goes over or under it. But if it's vertical to the ground, it becomes a wall. The side of a hoop house has a lot of vertical dimension to it.

Our farm is among the many that woke up one morning to see a mobile hoop house ripped to shreds and catapulted across the field. A squat A-frame Millennium Feathernet for the laying hens has both an extremely low center of gravity and no vertical impermeable walls. Because the roof is slanted, wind tends to push it into the ground rather than lift it up or knock it over. Again, please, please refer to *Polyface Designs* for the diagrams and specifications on these simple structures. Few things are as frustrating as dreaming up the infrastructure, building it, putting the animals in amid great fanfare and personal satisfaction, then waking up the next morning only to see everything strewn around the pasture. We call that a bummer, and yes, I've had my share of these experiences, which is why I'm going into detail here to spare you the same frustration.

Shade cloth is just as effective for summer shelter as an opaque roof, but without the wind issues. This means we can build lightweight shade structures for everything except chickens and not have to worry about wind. That's a game changer.

The next consideration for summer shade is whether to put it on skids or wheels. Skids can be wooden poles or steel pipe. The beauty of skids is that the structure can't roll and the center of gravity can be lower, making the whole thing less susceptible to wind damage. And an added plus: you can't have a flat tire if your structure is on skids.

The disadvantage of skids is that it's much harder to move. Any given structure will take more oomph to move on skids compared to tires. In general, I prefer skids to wheels. Normally, you're only moving animals a little distance, from paddock to paddock, and a skid structure is fine. A wheeled shade structure for big stock like cows can be pushed around and head off down the hill where it crashes into a tree and tips over into the creek. Guess how I know that?

Skidded shade structures are safer and more simple as long as you're not moving great distances or going up a public road. If you think you're going to haul your skid structure up a public road, make it narrow enough to comply with wide load requirements. Remember that it's always better to build two smaller ones hooked together like a train than one humongous one.

Small units are far more maneuverable. Goodness, you could invite some friends over and pick up a fairly heavy skid structure by hand if you needed to get it on a trailer for a big move. A big unit may impress your ego, but it's far less flexible than two smaller ones hooked together. Everything gets more complicated with large structures. Grow by duplication, not centralization.

If you do decide to use wheels, you can lower the center of gravity by nesting your structure between the wheels rather than having the wheels underneath. Outboard tires may look a little weird, but that adds a lot of stability to your mobile structure.

As you think about shade shelters for everything except chickens, consider multi-use. Certainly a shelter for pigs does not have to be as high as one for cows, but it's worth building a multi-purpose unit just for the convenience and versatility of interchangeability. Our shade cloth shelters for pigs and turkeys can be used for sheep and cows. One build can serve numerous functions and on a diversified homestead, that kind of multi-service option is far more valuable than the few dollars you might save building custom-fit shade structures for each species. Go ahead and spring for the taller version even if you don't have cows yet. Remember, using shade cloth immunizes the structure against wind damage.

Another reason to make mobile shade structures able to accommodate all species is because you may run a menagerie. At homestead scale, the most efficient way to run various species is in one group. Certainly all the herbivores could be run together in an electrified sheep netting. With a handful of animals, this option is often better than multiple groups, even if the netting is overkill for the cows. If you only have a couple of cows, you're better off running them altogether than separately. Do a time and motion analysis and you'll see why.

Every group, regardless of size, needs a fence to control it, a water trough, and shade. Non-duplication of effort where you can amalgamate always saves time and energy. The more compatible the species, the easier amalgamation is. At small scale, a menagerie is often the most efficient option.

In some cases, you may dispense with lanes and permanent fencing altogether and run multi species in one big electrified net or multi-strand mobile electric fence. In that case, a multi-purpose shade mobile is a must.

Now a word about tarps, used billboards and other non-metallic materials for impermeable roofs. I don't like 'em. Yes, they can be quick and cheap, but they have different convection principles. Have you ever been under a tent on a hot day? It's insufferable. Plastic and canvas don't reflect heat; they absorb it. That principle creates a thermal magnet rather than thermal diversion. Furthermore, all it takes is a loose, flappy piece and the next thing you know the tarp is in the next county.

Plastic and tarps only work on hoop structures because you can stretch them down hard. Plastic coverings only stay moored if they don't move. Securing them that solidly on a flat or non-arched surface is well-nigh impossible. The curvature gives something solid to stretch against. You don't get that leverage on a flat surface Even the Denver airport figured that out with its Rocky Mountain peaked stretch material roof.

That brings me to the idea of two-way trust. Part of animal happiness is providing a disciplined routine. In the big scheme of things, we want our animals to trust us. That's a big word. Remember, our animals always live in the moment, but they do have memories. That means they can't process unhappiness. They can't appreciate that we had a flat tire or that a friend died or we had an unexpected visitor that made us late. Our routine creates expectation and how perfect we are at fulfilling that expectation is where we'll be on the trust scale.

If you think people get antsy with unfulfilled expectations, triple it for animals. They will wait contentedly if they have a high level of trust, but if you're a minute late, they start into stress. This means that on your homestead you need to establish a routine for chores that doesn't change by more than a few minutes every day of the year. If you go away, times need to be part of the farm-sitter's requirements. It can't just be "sometime this morning milk the cow."

If you milk at 7 a.m., then milk at 7 a.m. Not 8 a.m. Not 9 a.m. Not 6 a.m. The cow is used to 7 a.m. Every time you alter a routine, the animals notice and that jeopardizes trust. We try to move our different herds of cows at the same time every day. Some are on a morning routine and others on an afternoon routine. They're glad for whatever schedule works for you; just make sure you don't flip the schedule around on them. If they're the 4 p.m. herd, then move them as close to 4 p.m. as possible, every day.

This is why at our farm, chores start with the chickens. They are the first animals up in the morning. Cows are second and pigs are last. When those chickens wake up and go over to an empty feeder, they get stressed. Minimize that by showing up on time at daybreak; that'll make some content chickens. I don't know if any research has been done on this issue, but I'm convinced it's a factor in health and productivity. When people complain about having trouble with their animals, often I learn that they have an inconsistent chore schedule.

I'm belaboring this point because "Oh, well, whatever" doesn't cut it with animals. They can't roll with the punches because they can't understand your apologies or explanations. All they know is the feeder is empty. This is nowhere more evident than in a herd of thirsty cows. Perhaps a water malfunction kept them from getting a drink for most of the day. You show up to move them at 4 p.m. and they're visibly uncomfortable and have the water trough tipped upside down. Maybe they even broke the valve.

You find the problem and get water flowing. The cows can stand there and see the water coming in. You're there. If they could reason like humans, they would know that all is well and simply wait their turn. Humans are the only creatures willing to wait in line. Animals don't. In spite of your calming reassurances that water is coming-"see it coming in?"--the cows mob, shove, bully, butt and fight each other to hog the water as it comes into the trough.

Human logic would say that you could leave them and they would quietly wait their turn, a couple at a time, like a line at a water fountain. But no, you must stay on site, sitting on the edge of the water trough, batting away the most aggressive, limiting access, for an hour until all have drunk, down to the smallest calf. Then and only then can you walk away knowing they won't drain it and knock it over again. Did someone say animals live in the moment?

Thirst, hunger, and discomfort all discredit your trust. In turn, this makes your animals harder to control, harder to manage. Anyone who has animals knows that each has a personality and some are just rogues. I asked Bud Williams, perhaps one of the greatest animal herders of all time, if he ever had an animal he couldn't handle. "Yes, I've had a couple cows we ended up shooting," he replied. Even the best don't win every time.

Bud's primary method involved building trust. One time he single-handedly rounded up and brought into a village a thousand reindeer. Over the course of more than a week he lived with them and established trust. Every time they would run away, he'd stop until they stopped. They learned that he would not chase them and gradually developed a trust. Any normal person would view what he could do as magic, but it really only involved trust.

Anything you can do to establish routine is beneficial. If you're moving animals across a field, make their cross fences open on the same side of the field every day. Don't flip from the east side to the west side. Feed at the same time. Gather eggs at the same time. Milk at the same time. While to a new homesteader this may sound like the very institutional rigidity you're trying to escape, it's actually the foundation of good relationships with your stock. If you cheat on this, your animals will be discombobulated and so will you. A routine is golden.

One step beyond trust is discipline. An uncontrollable animal can't settle and can't be content. The animal that's always out is not cute. It's a malcontent that will eventually lead others astray. Either get rid of that animal or get it under control. That doesn't mean abusing the animal, but it does mean doing whatever it takes to let the animal know it can't do whatever it wants to do.

An animal out of control is just as bad as a child out of control. Sometimes if you've let your management down and let a group of animals develop bad habits (like pigs getting out) you have to bring them in for retraining. If you're consistent and diligent, they'll fall into line pretty fast. Nearly 100 percent of the time, any bad habits in your animals merely reflect your bad habits . . . including negligence. I've learned this the hard way. Have we had some harder animals to train than others? Of course, but generally a roguish animal or group of animals are merely opportunists enabled by my negligence. If you have a problem animal, that's the way to bet.

The animals are simply doing what comes naturally to them. You can't blame them for going through an electric fence if you haven't kept the spark up. You can't blame them for tipping over the feeder if you haven't set it up right or let it run out. You can't blame them for eating their eggs when you haven't kept oyster shells in front of them. You can't blame the chickens for cannibalism if you haven't met their emotional and dietary needs.

All of this includes your pets. Lots of new homesteaders bring their dog or cat along with them, of course, and suddenly these lounging pets become vicious killers of chicks and lambs. Nothing stresses domestic livestock like being chased. These are all prey animals and they all still have plenty of instinct in them to know about predators. I've trained pet cats not to touch a chick until I say it's okay. They sit there and tremble, looking at the dead chick, until I say it's okay. Animal welfare cultists might not like the way I trained them, but this training empowered them to accompany me to the brooder every morning. I enjoyed the companionship and lots of times they got a dead chick for breakfast. Had they been allowed to attack live chicks, I wouldn't have been able to let them go to the brooder with me. Then I wouldn't be able to enjoy the company of the cat and the cat wouldn't have enjoyed the simplicity of an easy meal. Discipline is a good thing.

Realize that on our farm we raise many thousands of chicks. Anytime you have thousands of animals, you have dead ones routinely. It comes with scale and the law of averages.

Despite all your efforts in trust and discipline, it all breaks down sometimes if a mama feels her baby is threatened. Probably the only time in my life that I thought a cow was going to kill me involved a misstep grabbing a baby calf. This was back in the days before we moved our calving schedule to later in the spring and the cows were still on hay in the hay shed when calving started.

The sacrifice area was muddy and I wanted to ear tag a calf. I knew this particular mama cow was highly protective so I waited for my chance. I fed the cows and when they went in the hay shed to eat, I snatched the sleeping calf. My plan was to grab it, step over the electric fence away from the mother, tag it, and return it to her, all in a few seconds. The plan derailed when I slipped in the mud, carrying the calf, and fell forward on top of it. It bleated.

Folks, you cannot imagine how fast an upset mama can move when her baby is in danger. Before I could even get up from falling, that cow vaulted out of the hay feeder and was on me, bellowing sadistically and pushing me into the mud with her forehead. I screamed, of course, terrified, and tried to get away. But every time I tried to get up she smashed me back down, stepping all over my legs and literally driving me into the mud.

Finally I was able to crawl far enough away that she took a breather and let me roll out under the electric fence. She immediately went over to her calf and left me to my bruises. Fortunately she did not step on my head or torso; only my legs. I partly crawled and partly limped to the house where I undressed and took a hot shower. Teresa pulled a handful of rocks out of my pants pockets that came in as she ground me around in the mud. We decided the mud saved me from broken bones because when she stepped on my legs, they could mush down. It looked like I'd been through a gauntlet; my legs were black and blue for days.

A balance exists between docile and aggressive. You want a mama who will protect her baby, but a mama who thinks nothing of killing you might be a bit over the top. We kept that cow. When I heard about neighbors losing calves to coyotes (to my knowledge, we've never lost one to coyotes) I'd always grin and say "you don't know about Number 35." I always wished a coyote would try to attack that cow's calf; end of coyote.

I was about 6 years old when I first encountered maternal instincts that scared me to death. My dad, brother and I had gone to a farmer a few miles away to buy a couple of weaner pigs. The farmer had not yet weaned them, but they were old enough and big enough to wean. It was dark and we had brought a gunny sack to put them in and bring them home.

We all stood outside the pen and the farmer went in to get the two piggies. This sow looked completely comatose, lying there with her piggies nestled up next to her. The farmer eased up close and grabbed two of them. Mind you, these piggies were no bigger than a good-sized cat. They weren't heavy or awkward. He was literally two steps away from the pen's fence and safety, but that sow responded faster than lightening when the first piggy squealed. That's all it took. I'll never forget how 550 pounds of pure hog wrath erupted off that floor. She caught that guy's pants leg in her mouth first. The whole pen was only about 10 feet square; he could have almost reached in over the side and grabbed the piggies, but they were a handbreadth out of reach.

He threw the pigs to my dad as he collapsed against the pen gate. The sow got another bite on his leg as he flung himself out over the pen. I was transfixed. Everything had been calm and beautiful one moment--what can be more sweet than a sow nestled in with her 8 piggies? But in a moment, in the twinkling of an eye, everything turned to chaos and I remember thinking in that split second that the sow was going to literally kill that farmer. That was nearly 60 years ago but I still remember it like yesterday, so powerful was the moment.

A dear, dear friend of ours was nearly mauled to death by "the sweetest, gentlest sow in the world" when she inadvertently stepped on a piggy and the sow attacked her. The sow nearly ate off one arm, broke ribs, punctured lungs--our dear friend barely escaped with her life. So folks, I'm not trying to scare you; I'm telling you that all the trust and all the discipline in the world sometimes cannot overcome mama's wratb. Don't mess around with this.

A proper respect for what a crazed animal can do might save your life. Fortunately, these incidents are few and far between, and they're the stuff of stories and special memories. But never think that an animal has no mind and isn't thinking. Remember, they're in the moment. They don't want your explanations and they don't want your reasons. All they want is to protect their babies. Respect that; be wise; be careful. Don't get yourself in a compromised situation.

Most of this book is about the joy, happiness, and abundant provision of homestead livestock. It's positive and upbeat. But I would be unfair if I didn't broach the subject of risks. Most old livestock farmers have battle scars. I sure do. But that's why you want trust and discipline and well-designed control mechanisms. These are all part of the healthy and happy experience.

I like Bud Williams' most common admonition: "Whatever you're doing, slow down." Generally, the most common solution to all human-animal frustrations is to just slow down. Sit down. Back off. Let the animals know you aren't going to bird-dog them every minute. Give them a chance to settle, to realize you aren't trying to squeeze them forever.

Here at Polyface we run a professional and formal apprenticeship program and the natural intrepidness of non-farm folks is amazing to those of us who spend our lives with beasts that can maul us. You never turn your back on a bull, no matter how gentle he is. We don't want to teach fear, but we do teach awareness of the gravity of the situation. Always have escape routes. Get off your cell phone. Take out the ear buds. Be 100 percent aware of the situation. It can change in a heartbeat.

An animal becomes unhappy for a reason. Occasionally animals are mean tempered. I'll invoke the Biblical injunction here from Leviticus 21:29 that if an ox had a bully temperament and eventually kills someone because the owner didn't keep him confined, then both the ox and the owner should be put to death. That's a strong injunction, but it shows that difficult animals have been around for a long time and their owners are ultimately responsible.

Do your best, and if it's not good enough, get rid of the animal; that's my advice. Life is too short to try to reform dangerous animals. That includes biting dogs and butting horses. This book is about happiness, not horror. I'll leave the sobering admonitions there and hope this word to the wise is sufficient.

Chapter 15

Irrigation

Gardening books are full of irrigation information, but they don't extend it much beyond intensive vegetable and fruit production. That's a mistake. Our homesteads and commercial farms should be viewed as giant gardens. Viewing them otherwise gives us license to be complacent about water management.

Irrigation is an underutilized tool in our forage management portfolio. We've talked about soil fertility, animal movement, grass growth and even pasture cropping. Let's focus on irrigation as a viable option to reduce valleys in the forage chain.

Whether climate is becoming more erratic is not the issue. The fact is that compared to temperature, water is a far greater limiting factor. Like anything we manage, the biggest results come when we attack the weakest link. Water limitations are not just an issue in deserts; they are everywhere.

Not long ago I was in the Netherlands when that region suffered a crippling drought. The Netherlands! Even a country half submerged in water suffers periodic droughts. There, two months without rain are equal to Utah going two years without rain. Drought impact is based on what is normal. Local moisture determines the type of vegetation and the type of farming. It's all relative.

Plants and farming systems become acclimated and adapted to a certain climate pattern. When it fails or stresses, everything in the ecosystem responds negatively. I won't repeat the material in the water chapter except to reiterate two rules developed by P.A. Yeomans, the Australian water guru who developed the keyline system: try to eliminate surface runoff and never end a drought with a full pond.

Excess water should be stored and then it should be utilized in times of moisture stress. Most of us suffer perception damage regarding irrigation, thinking of it only in terms of massive pivot wheels and huge water cannons. But simple, low cost systems have been developed by New Zealand graziers that have wonderful applications world wide.

Like all the ideas in this book, the beauty of small holdings is that you don't need industrial-scaled management and infrastructure to stimulate high production. Production data offers incontrovertible evidence that even novice backyard gardens produce more per square foot than the most advanced techno-sophisticated commercial operations.

When you hand spade sweet potatoes, for example, you can beat the bruise-and-scar percentage by long shot compared to machine harvesting. Walking behind a zucchini-picking crew in Colorado one time, I was appalled at the waste. A missed squash simply had to sit there because the harvesting machine couldn't be slowed down. On a homestead, you can be far more meticulous in all aspects of planting, harvesting, and processing.

Irrigation is similar. We're not talking about 2 ft. diameter wells pumping 200 gallons a minute for almond groves in California. We're talking about a few acres, hopefully not in the desert. Certainly some folks will want to homestead in the desert, but most will be in more temperate areas where natural rains do the heavy lifting and irrigation acts as a supplement, rather than the main engine. If you happen to be in extremely arid areas, though, be assured that many procedures can leverage the rain that falls. Permaculture is the teacher for this.

On our farm, once we had enough water in our savings account (ponds) to think about irrigation, we did some poor-boy trials to see what it would do. In our 31-inch rainfall area (Lancaster, PA and Cherokee, NC are both 52, just for reference) we usually have a dry time even if not a full-blown drought. Any time soil moisture becomes a limiting factor, it reduces vegetative growth. Plants need moisture in order to thrive; some of course are more tolerant of dry conditions than others, but all will eventually die if they don't get moisture.

Being frugal bootstrappers, we decided to do some experiments with off-the-shelf stuff. We bought a 2-inch output, 50-gallon-per-minute $200 pump from Tractor Supply and a couple rolls of collapsible pipe for supply lines. We hooked that onto a piece of black plastic pipe with Ts every 25 feet to accommodate garden hoses running out to some little $10 whirly-gig garden sprinklers. The whole arrangement cost under $1,000.

The one big modification we made was switching out the gas tank on the pump to a 5-gallon gas can so it would run all night. While this all sounds clunky and crude to irrigation professionals, it allowed us to cover a 5-acre experimental field, track time and fuel, and monitor additional vegetation growth versus the rest of the field. For now, don't get bogged down in finding fault with using gas to spread water. This was an embryonic experiment, not the final model. The results were stunning.

The biggest cost, of course, was the initial pond construction. Remember, I don't believe in pumping out of aquifers or streams; the only ecologically ethical water source for irrigation, in my opinion, is from impounded surface run-off. That's inherently excess in the system that can be skimmed off into savings. Maintaining an ecological balance sheet is as important as maintaining a financial balance sheet. That pond cost about $3,000 to build.

The bottom line on the experiment was that for a total one-time investment of $750 per irrigated acre, we could double production in an average year. When I say average, I mean that at some point in the growing season water is a limiting factor in forage production. That means that if land costs $5,000 an acre, it's far more economical to invest in irrigation to increase production than to buy more land. It also means that you can take your small acreage and make it equivalent to a lot more, just like the well-managed grazing program.

Remember that if you have a 5 acre drainage in a 30-inch rainfall area, a third of that will come in quantities that the ecosystem can't absorb, resulting in surface runoff. One-third of 30 inches is 10 inches X 30,000 gallons per acre-inch is 300,000 gallons X 5 acres is 1.5 million gallons. In temperate areas, if you can get 6 inches of water in a dry time, it can literally grow an entire production cycle that would not have occurred otherwise.

As soil dries out, the vegetative growth slows. It doesn't continue growing well right up to the dryness brink and then suddenly stop. The growth decrease is a gentle curve. All irrigation gurus agree on one thing: never let the soil get bone dry. You get far more bang for the effort if you start applying some water earlier rather than later. When the soil gets bone dry, it takes a lot of time just to re-awaken the biology and get things going again. If you jump on it early and never let the soil get extremely dry, the results will be far better.

In the long run, you'll get more benefit from a given amount of water if it's applied before things get too dry. If you jump the gun and rains come back, your pond will refill. You'll never regret starting the irrigation remediation early, whether rain comes or whether it never does and you actually pump your pond dry and have drought stress anyway. Keeping things growing is much easier and more efficient than resurrecting them once they've stopped.

Just imagine that two years out of three, irrigation

capabilities allow you to add 100-cow-days to your pasture (that's about 3,000 pounds of dry matter). If that's worth $1.50 per day, you've increased your gross income by $150 per acre. That's just raw production income and does not measure increased manure, organic matter, or reduced fertility loss.

We initially looked at irrigation as a way to get water on the manure behind our broiler shelters. Watching all that manure bake in the hot summer sun was like watching fertility leak off the farm. Our first priority was to try to get all that chicken manure watered down and pushed into the soil before it evaporated into the cosmos.

Today, we don't use gasoline engines to power our irrigation and we don't use collapsible pipe and we don't use garden store sprinklers. Fortunately, the Kiwis in the last couple of decades developed the K-line system which was invented by and for graziers and not crop farmers. Although it will work in crop farming, it's especially conducive to pastures.

The K-line system utilizes a tough trunk line to supply sprinklers nested in what looks like a dog dish saddled to the trunk line. One trunk line can service up to 12 sprinkler nodes, each of which covers a 50 ft. diameter circle. That means a full-on 12-node unit can cover 600 feet X 50 ft., or 3/4ths of an acre with an inch of water every 8 hours. If you want to put on 2 inches per application, that would take 16 hours. In general, we move it twice a day for 1.5 inch applications covering 1.5 acres per day. In a week, just one of these units can cover 10.5 acres.

We use electric pumps on 220 power that run for pennies a day; pumping water does not take much energy. With four of these installations on strategically located ponds, we can cover a substantial amount of our farm with water insurance. If it takes roughly 10 minutes per day to move the line to a new area (you hook onto the trunk line with an ATV and pull it wherever you want), that's about an hour a week per 10 acres. What is your time worth? Let's say $20 per hour. That means our labor cost is about 6 minutes per application, or 1/10 of an hour, or $2 per application. Are you hanging with me here?

If we do four applications (6 inches) to get that additional 100 cow-days worth $150 over a 45-day period, for example, our labor bill is $8. Electricity is another $2 and depreciation is $2, meaning we've invested (not including the initial cost of pond construction) about $12 per acre for a return of $150. That sounds like a pretty decent return on investment to me.

Now how big a pond will this require? If we need 6 inches of water for 10 acres, that's 6 inches X 30,000 gallons per acre-inch is 180,000 X 10 acres is 1.8 million gallons. A cubic foot holds about 7.4 gallons of water, so our pond needs be a bit less than 250,000 cubic feet. What does that look like? It looks like a pond 185 feet across averaging 8 feet deep. That's a big pond that might cost $7-10,000. But in the big scheme of things, think about the things farmers spend that much money on without batting an eye. Machinery, trucks, bulls. Goodness, investing in resiliency wins hands down.

And realize that all of these costs and requirements on the development side of the equation vary based on size. In the above scenario, we assumed 10 acres, but if you're only irrigating 5 acres, everything drops by nearly half. One of the beauties of the K-line system is that it scales easily. In other words, getting a 5-spinkler system is about half a 10-sprinkler system. Certainly you can get volume discounts, but they aren't that big. One of the most attractive things to us, initially, was that we could get in on a shoestring and if we liked it, we could add sprinklers. We started with the homesteader start-up plan and then added components as we got more ponds built and electricity extended to these strategic ponds.

Sometimes you'll want to put your pumping pond close to your electricity but that might not be the most conducive spot to catch water. We have several installations where our big supply ponds are far away and gravity feed to the pond nearer our electric supply, where the pump is. When we're irrigating, we let these farther-away ponds flow into the small pumping ponds through a 2-inch plastic pipe, kind of like a big nurse tank supplying a working tank. If you can't get gravity flow, you might need to pump from a farther away pond, but if it's anywhere near the level of your pumping pond, you can move a lot of water extremely cheaply. If you're not lifting it much, water moves horizontally with little energy.

For example, we have two large ponds that service a couple of our small pumping ponds, but they are not quite high enough to gravity flow. We use a gas pump to refill the pumping pond; in one tank of gas it'll fill the pumping pond enough to supply irrigation for a day or two. The extra vegetation grown more than compensates for the petroleum used. The whole earth cycle is about matching vegetation growth to carbon use; as long as we're putting back more than we're using, I don't have any problem using carbon.

Irrigation is a way to bring human ingenuity to the landscape to make it capable of producing more biomass per acre than it could in a natural state. Even on a small homestead irrigation pays and can fundamentally change the productivity of a pasture. That is part of the human mandate toward stewardship.

With this modular technology, you can begin irrigating at tiny scale and increase incrementally. Too many irrigation systems require a $30,000 infrastructure investment and don't pay out until you're covering a hundred acres. Finding technologies that scale is rare; this K-line system is one of them.

I've focused most of this discussion around what is known as the summer slump. That's the time in mid-summer when things don't grow as well due to heat and lack of moisture. Sometimes, though, the dry limitations are in the spring or fall, or both. Not long ago we found ourselves irrigating in early May when the rains shut off like a water spigot the first week in April. It was downright scary.

We fired up the irrigation to make sure the hay would come on, and it did. As soon as we finished making hay, the heavens opened and we had April rains in June and July. That doesn't happen often, but a lost hay crop can be devastating. In recent years, we've used the irrigation as much in the fall as we have in the summer.

Fall is when we're trying to accumulate winter stockpiled forage. We love those open, warm falls, but if the ground gets dry, all that beautiful fall weather goes to waste. Interestingly, in our area the highest average rainfall of the year, historically, is August due to hurricane season. But this belies the problem with averages. One year a hurricane comes through and we get 15 inches. Then the two subsequent years nothing comes through and we only get 2 inches, or sometimes nothing.

On average, we get 5 inches of rain every August, but that's far from an accurate representation. The truth is we got flooded one year and burned up the other two. With some irrigation insurance, we can keep that deep production valley from occurring and know that we'll have good growth up until the temperature drops enough to stop it. As long as the temperature is warm enough to grow stuff, we want it to grow.

Seldom do people consider irrigating pasture, and even more seldom to small-holders consider this option. Now that we're doing it routinely on our farm, I'm a bigger fan than when I was just dreaming about it. Knowing that we can throw water on an area whenever we need to sure changes the worry and sleepless nights when the clouds refuse to gather.

Dry times are almost always during the growing season. You never get as much sunshine to grow things as you do in a drought. Heat and sunshine make it the perfect time to capitalize on those two critical factors for good biomass production. Irrigation makes it possible to capture those conditions every time they happen.

Chapter 16

Laying Hens

I can't imagine a homestead without chickens. I can't imagine a backyard without chickens. I'm not sure I can imagine a condo without chickens. They're the ultimate all-purpose partner on any self-reliance team.

They don't need heavy duty infrastructure; they're child friendly, and they turn life's nasty garbage into one of the most perfect foods in the world. What's not to love about chickens?

Plenty of breed books exist to explain differences in genetic traits. I'll leave that to the experts. Instead, I'm going to dig into the practical day-to-day issues common to all breeds of laying hen. Some breeds naturally lay more prolifically than others.

My bias is toward the traditional dual-purpose American varieties. They're called dual purpose because the cockerels can be used for meat and the hens lay fairly well. The tiny-bodied hybrids do only one thing well: lay eggs. But it comes at a cost: fragility, flightiness, failed smarts and not much meat on a carcass.

These hyped-up hybrid layers don't learn as fast, like how to come in at night. They're less hardy to handle inclement weather and diseases. In general, they're less docile. That means they get stressed easier, which translates into egg anomalies like blood spots. I've not seen research on this, but it seems realistic that a larger-bodied hen laying 10 or 20 fewer eggs a year than these hybrids will have more energy to put into an egg than a bird that weighs half as much and lays more prolifically.

The heavier bird carries some reserves and that makes her more functional for a homestead. Here at Polyface, we now hatch most of our laying hens and hope to hatch them all in the future. We select the oldest birds each year; what we call survivor genetics, for our breeders. Over the last several years, we've seen remarkable progress in these birds in all the characteristics that matter most: docility, hardiness, body size, and laying longevity.

My first entrepreneurial farming foray was a box of 50 as-hatched heavy breed specials from Sears and Roebuck in 1967. I still remember the wonder and awe I enjoyed that morning taking those chicks out of the shipping box and putting them into my little cardboard brooder box under a heat lamp in the basement. Few things can compare to new life.

As is typical in those instances, my "as hatched" 50 chicks ended up being 18 pullets and 32 roosters. Don't get me started. Anybody who gets "as hatched" laying hens with equal males and females, please tell me and I'll put a plaque on the wall with your name on it. Hatcheries notoriously use these schemes to move their cockerel overages.

Prior to the 1960s, young meat chickens called broilers came from the males of these dual-purpose chickens. That meant that the boys had some value. But as the broiler industry developed the fast-growing double-breasted Cornish cross, this historical market and use ended, eliminating any value for these cockerels. Once that happened, the only use for these historic dual purpose birds was as layers. Of course, the industry went to smaller-bodied hybrids; the divergence in the poultry industry between the eggs and broilers was profound and gradually developed into the specialized industry we have today.

As we hatch more and more of our own birds, we struggle with the same issues within the industry: what to do with the cockerels. We raise some and sell them as heritage birds to a small but loyal cadre of gourmand customers. Most now go into dog food. Until we found the dog food outlet, we threw some away too. But at least for now that has stopped. The advantage of a homestead, of course, is that you can eat these cockerels in your household and honor the brothers as well as the sisters.

The industry still throws away millions of the dual purpose cockerels and that's a shame. At least the hens are worth eating when they're done. We'll circle back to spent hens later.

Let's start with the lay cycle. The hen's lay ability is affected by three things: age, day length, and comfort. Normally you'll start with a day-old chick that begins laying around her 20th week. If it's spring and the days are getting longer, sometimes she'll start laying as early as week 18, and if it's fall with declining day length, she might hold off until week 22 or even 23 to show you a first egg.

Once she starts laying, she'll lay well for most of a year and then molt. She'll drop a lot of feathers, get new ones, and lay virtually nothing during that time. It's a time of renewal for her; letting her go through the process is important to maintain health and longevity. The second year she will lay within 10 percent of her first year, and all her eggs will be large.

At the beginning, a pullet lays introductory pullet eggs, which tend to be smalls and mediums. As she matures, the egg size increases. If the feed is too high in protein, the eggs might get too big. On a homestead where the birds get lots of kitchen scraps and garden wastes, they adjust their diet to what they need. I take great stock in a chicken knowing all she needs to know to be a successful chicken. Since she doesn't watch TV, she's not swayed by fancy advertisements or what celebrities tell her to do.

Again, in this book I'm not going to dig into the minutiae of brooding and rations. Very seldom will a homestead flock getting lots of scraps be unable to balance any ration shortfalls. They're pretty resourceful at finding what they need.

A hen will lay beyond two years, but it will taper off. The birds we're hatching seem to lay extremely well for three years and that's a game changer. *The Small Scale Poultry Flock* by Harvey Ussery is probably my favorite of all the homestead-scale laying hen books and addresses many of the issue's I'm going to purposely leave out here. No sense re-inventing the wheel. With his encouragement, you may want to try hatching your own chicks under a hen. I can assure you that these on-site hatched pullets out-compete their commercial hatchery cousins as a general rule. But most people, including us, still rely on hatcheries even if we do incubate many of our own. Patronizing independent hatcheries is a gift to national food security from small-scale chicken lovers.

As a hen ages, her laying percentage drops off and she's more and more susceptible to sickness. As a layer, lots of things are going on inside. I wouldn't keep a laying hen more than four years even if she looked healthy. Why complicate your life with geriatric chickens? Go ahead and cull her while she's healthy and put her in the stewpot. That makes room for new birds, keeping your flock fresh and productive.

As a bird begins to lay, she begins robbing her body's yellow pigment to put in the egg yolk. This pigment comes out of her body in the following order: vent, eye ring, ear lobe, beak, feet, and shanks. When she goes OUT of production, it comes back in the same order. So if a bird is bleached around her vent and eye ring, but has yellow pigment on her ear lobe, beak, feet, and shanks, you know she's coming into production. If she has yellow pigment around her vent, eye ring, and ear lobe but is bleached in her beak, feet, and shanks, you'll know she's going out of production.

Pigment is the most important giveaway in how well a bird is laying. The next indicator is pliability of the public bones and abdomen and the space between the pubic bones. Always hold a chicken with your middle finger between her legs so her breast fits into your palm and her head is toward your body, rear pointing away. This procedure protects you from being pooped on but more importantly, it's comfortable for the chicken. Her legs dangle down between your middle finger and your ring and index fingers. If she starts to peddle, her feet don't scratch you; they just peddle out in the air. See page 88 for how to hold a chicken.

This holding technique allows you to examine the bird all over and even flip her upside down without stressing her or getting scratched. Holding the bird this way, take your free hand and push gently around her vent area. You'll feel two pointed bones beneath the skin; those are her pubic bones. When she's laying well, the daily stretching and exercise widen the pubic bones enough to put two fingers in between them. When she's not laying well, you can barely get one finger in between them and they become hard to the touch. Soft and pliable is what you're looking for in the abdominal area.

With this technique, you can tell which birds are laying and which ones aren't. In a commercial flock like ours, we generally do an all in, all out approach. When the flock ages out, we process the whole group even though a few birds are still laying well. That's the inherent inefficiency and waste of larger scale operations. But if you only have a couple dozen hens, you can cull individually.

If you have a group of new pullets you want to introduce to an older flock, here is our protocol. First, realize that chickens are probably the biggest bullies on your farm. Worse than cows, sheep, goats and even pigs. Chickens will actually kill each other to gain ascendency in the hierarchy. Chickens establish their social order, also known as the pecking order, through both psychological and physical intimidation. The ones on the top of the pecking order despise competition or any change of circumstances. I guess they're kind of like politicians.

Never toss a group of new birds into a group of old birds without making some accommodations. Here is a protocol for safe assimilation. Ideally, you keep the new ones and old ones separate, but allow them to see each other. That way they get acquainted. When you take the partition down (or open up the door in it) leave the second feeders, waterers, and nest boxes; don't consolidate them.

If you can't put a partition between them, then make sure you put secondary feeders and waterers on opposite ends of the housing area. The old hens will aggressively chase the new ones away from feed and water that the old ones are used to. If you put in additional feeding and watering stations, though, on opposite ends of the area, the old hens will be hard pressed to deny access to the new stations that far apart. They'll run themselves to death and eventually give up the chase. Don't move the old feeders and waterers; leave them in place. The older hens will stay with those in that area, to guard them. A new set of feeders and waterers, at the other end, will accommodate the new birds and grant precious relief from the old hen bullying.

The main thing is to make sure the new birds have everything they need without assimilating with the old birds. Let them live together but segregated. Gradually they'll get used to each other and full assimilation happens within a month. It certainly doesn't happen quickly and you might get frustrated at the continued territorial protection after several days. But be patient. They'll assimilate eventually.

To roost or not to roost, that is the question. Decades ago, thinking chickens needed roosts, I constructed beautiful commodious roosts and then was chagrined to see only half the birds using the roost. I asked an old timer neighbor about it and he said "well, you have to train them to go to the roost." I decided then and there that roost requirements are overrated. I appreciate that in the wild self-preservation demands roosting up in trees to get away from ground predators, but in protected places, the birds are just as happy resting on the ground.

That may get me crucified by some folks, but I figure if I have to teach them to get up on a roost, they must not need it to fully express their chickenness. We do have roosts in our Millennium Feathernets and some of the birds use them but some don't. We see the roosts as a way to create more vertical resting space, to use cubic space rather than just square footage space. The roosts, in this case, are built into the structure and don't complicate floor space or head space. In this instance, the roosts serve a space purpose, not an animal welfare purpose.

Culled laying hens, known as stewing hens, have wonderful yellow fat and exceptional taste. Our favorite recipe for them is to cook as many as we can cram into our biggest roasting pan, set the oven on 350 degrees F for 3-4 hours, and then pick the meat off. You can tell when they're done by the way the skin pulls away from the legs. You want the legs to fall away from the body a bit, stretching the skin so it pulls away from the body of the chicken.

After letting them cool, we pick the meat off, chop it into bite sized pieces, and pack it in freezer containers. That way we have pre-cooked deboned chicken ready for casseroles and meat salads. Compared to freezing whole birds with all that airspace in the body cavity, this procedure compresses valuable freezer space.

That's the life cycle of the laying hen. Now let's look at the next big factor in laying: day length. Birds respond to increasing day length and lay accordingly. They also respond to decreasing day length, and reduce their laying accordingly. If you think about the natural cycle of a chicken, you would have extra eggs to hatch in the spring. From late summer until the following spring, you need all the eggs for eating; historically, the only time folks had enough extra eggs to set aside some for hatching was in the spring.

A bird that hatched in late May, for example, would begin laying in early November. A bird that hatched in early April would begin laying in early September. The point is that in the natural scheme of things, new pullets joined the homestead flock during decreasing day length. The beauty of that is that when the bird begins laying, her age dictates that she increase her production. As she develops into a full-fledged layer, her natural production trajectory goes up. In the fall, that's directly counter to what the day length is telling her body to do.

This brings up the important point that these three factors--age, day length, and comfort-- need to be managed so they don't all convene at once. Ideally you only have one of these strikes; two is serious, and all three at once is devastating. The whole goal, then is to spread these three strikes out during the season.

Starting pullets to lay in declining day length is a way to make the life cycle run opposite the light cycle and minimize laying rate reduction. Some people put lights on the birds in the fall. Lots of nifty little solar and LED options exist now, even for Eggmobiles, that didn't exist a few years ago. I don't have a problem putting lights on them in the fall as long as they are withdrawn by the winter solstice: Dec. 21. That's the shortest day of the year. The summer equinox, June 21, is the longest day of the year.

Unless you put lights on the birds, those two dates exercise a lot of control over the lay rate of your flock. Birds need a period of rest and recuperation some time in the year. If you don't give it to them, they'll be more prone to sickness. Give your ladies some time off. The easiest way to do that is to start a new batch of pullets every year to begin laying in August or even September so their life cycle biology makes them want to lay well in December. No matter what you do, they'll still perk up after Dec. 21, but a healthy young bird that is comfortable will hang in there at 50 percent or better without lights even in December. All of these dates, of course, would be reversed in the southern hemisphere.

Stagger your new pullets so that you have some young birds and some old birds. The old birds, during the winter short days, will rest while the young birds hold up production. In the spring they will all crank into high gear.

That brings us to comfort. How do you keep a laying hen comfortable in a blizzard? Remember the one thing a chicken detests: getting wet, especially cold wet. Like most animals, chickens can handle virtually any cold as long as they are dry and aren't in a draft. If the air is still and all their feathers are dry, they can handle extreme cold. But they won't lay.

The reason is that the bird has to use up calories to maintain her body heat. If she uses up all her energy to maintain her living temperature, she doesn't have enough energy left over to make an egg. A bird's metabolism is high anyway; dumping discomfort on her uses up any energy reserves. A chicken's comfort zone is anywhere from 40-80 degrees F.

Chickens don't like extreme cold or heat. On a hot day, they'll find shade the best they can and hunker down for the greatest part of the day. Early morning and early evening they'll venture far and wide, but on a hot, sultry day they don't want to be out in the hot sun. A roof on a portable structure works great for warm season comfort.

But in the winter, more discomfort elements exist. Hoop houses work extremely well to bring both light and comfort to the birds when the sun is up. At night they get cold, but at night they aren't moving around anyway. They just huddle together and sleep. It's kind of like a cold bedroom; we just snuggle under the covers and we're fine. What we don't want is a cold bathroom, kitchen, office, or work room. Freezing to death while you're trying to do things is no fun.

Hoop houses offer daytime heat to keep the chickens warm when they're up and working. The cold at night doesn't bother them too much. Our specifications and designs are all in *Polyface Designs* so I won't repeat them here. The bottom line is that if you don't make your chickens comfortable in the cold, they won't lay any eggs. Getting them warm during the working day is critical to getting eggs in the wintertime.

One of the hardest multi-season pieces of infrastructure to make is an all-weather shelter for chickens. In the summer you want something airy that lets them maximize their time on pasture and keeps them cool. In the winter you want something warm and tight that minimizes mud and cold, damp conditions. Those are the conditions that encourage sickness. These two protocols are extremely hard to combine in one structure.

Many years ago before we built hoop houses, I tried all-season housing using our Eggmobiles. I covered the slatted floor with sheets of plywood and put in a bunch of old hay to absorb the manure and give them something soft to walk on. Those poor chickens were unhappy. They didn't die, but they just huddled around forlorn all day and definitely did not lay any eggs. I could have put plastic on the air space below the roof to tighten it down, but in the end, it was like living in a metal can in the freezing cold. Everything sucked the heat out of those poor chickens.

You could make a portable hoop house and cover it with shade cloth for the summer, but then you have wind issues like I've described previously. In the winter, you'll have a dickens of a time putting in deep bedding for them because it covers up all the bracing. It'll come up against the walls and the birds will be able to reach the plastic and peck it into oblivion. Short of moving to a place that doesn't get extremely cold in the winter, I haven't figured out a structure that works equally well in both hot and cold times.

To be sure, if you had only a handful of chickens, you could certainly have a hoop house you would continue to move in the winter, but if you get a blizzard and have to let it sit a week, even a few birds will muck up where they are and then they'll get wet, muddy feathers. You get the picture. Fortunately, chicken infrastructure is by far the cheapest and lightest to build so customizing winter and summer quarters is not too bad. Furthermore, I don't recommend any portable hoop structures tall enough to walk in. They're too susceptible to wind damage.

When the winter winds howl, I sure don't want my hens in a flimsy mobile hoop house unattached to the ground.

That said, the one set-up that can be all-season for chickens is a deep litter system that does not include free ranging. Loose housed birds in an all-weather structure, with windows that can be buttoned up for winter and opened for summer breezes, on deep bedding, can work just fine. A hoop house top with shade cloth covering in the summer (over the plastic) and removed in the winter is certainly an easy way to make sure the birds can get warm in the winter. This is a non-portable structure using deep bedding.

The Achilles Heel of chickens is predators. When hosting a farm tour at our farm, standing by the chickens in the field, I ask for a show of hands for who likes to eat chicken. Invariably, every hand goes up. Then I say: "Everything out here that goes bump in the night likes chicken too." Plan a fortress for your chickens' night time abode.

Predators are opportunists and look for easy pickings. If it's too complicated, they move on to easier targets. Over the years, our number one predator has been rats. Fortunately, they won't take a chicken after a couple of weeks old, but they'll sure clean out your chicks if they can get in. A rat doesn't need much room. One inch is plenty of space for a rat to enter. Make sure your chick brooder is rat tight. A couple of aggressive cats will serve your homestead well.

The next most common predator is a raccoon. They especially like the chicken head and breast and are extremely creative at getting in. Fortunately they don't usually kill for the sport of it. They kill what they want to eat and then leave.

A possum likes the rear end of the chicken and will routinely yank out all the intestines. Possums delight in killing for the fun of it and will wipe out a whole flock if given the chance. If you find guts strewn all around, you can be pretty sure it's a possum.

A weasel or pine martin likes blood. If you can't find any markings on the dead chickens, pull back the feathers around the neck and look closely. You'll see two teeth marks, like a vampire, in the skin where these pesky critters sucked out the blood.

Both foxes and coyotes are wily and don't like coming around where human scent is strong. To my knowledge, we've never had a coyote attack our chickens and we've only had a couple of fox attacks. The worst losses to foxes are when a vixen has a litter of pups. The distinctive sign of a fox is that it tends to carry the bird to its den and eat it there. You won't see any feathers; all you'll notice is that a couple of chickens are missing. If that's the case, chances are you've got a fox visiting.

Dogs, including the neighbors' pet dogs, are certainly an ongoing problem, especially if you live in a highly populated area. We've had problems with dogs primarily on leased property next to houses and lots. We rent one place bordered by 22 houses; that's where we've had by far the most dog issues. Dogs kill for the fun of it. If you go out and see what looks like a bomb went off with dead chickens lying around torn up, not eaten, haphazard, you can be pretty sure you've got a dog on your hands. It's probably not a wild dog; it's probably a neighbor's dog.

Fortunately on small holdings, you're never that far from your animals so the chances are you'll see dogs if they come. You can always mount trail cameras if you have something picking your pocket but you can't figure out what it is. I talk about predation extensively in *Pastured Poultry Profits*; be assured they are a real problem and the main deterrent is a tight night structure. Most of these attacks are at night; rarely in the day.

Because predators are opportunists with a short attention span, they generally don't spend a lot of time in one spot. They poke a bit, dig some, rip for a minute, and if things don't open up quickly they head on to another place. This is why mobile shelters deter more than most people realize; each day the predator has to start over. But a stationary shelter offers ongoing opportunities for these bursts of interest. What would be a daunting dig in one night, for example, isn't daunting in short 15-minute bursts over a week of nights. A stationary structure must withstand repeated assaults that can gradually accumulate to a breach.

The most common daytime predator is something aerial. All I'll say is do what you have to do. I recommend a guard goose with your flock. But some are aggressive and some are wimps. You'll have to let the goose tell you which it is. Well trained homestead dogs are a treasure. If they're on the ball and not too old, they'll patrol your premises at night and if they can't fight the imposter off, they'll at least let you know about a breach in the homestead perimeter. Shock collars work well to train them.

This whole predator issue brings up the question of free range versus totally enclosed systems. The evidence is pretty clear that the bigger the flock, the less likely an aerial predator--or any predator for that matter--will attack. Predators are a bit intimidated by large flocks, which I would say are anything more than 500 in a group. This puts the homestead flock at a huge disadvantage in predator susceptibility.

If you're letting your birds free range without protection, the question is not if but when you'll lose some to hawks or other critters. It's inevitable. For small flocks, I'm a believer in complete enclosure. That's roof and sides. Floorless portable shelters that keep the birds constantly enclosed can offer fresh pasture every day but total protection. Let's think this through.

We'll compare two systems with 24 birds; one is totally free range and the other is totally enclosed in a mobile coop. The free range birds need a visit in the morning to drop their nest boxes, check on them, and open their access door. Or perhaps you surround them with electrified poultry netting; that's fine, but you'll have to move that every few days so whether you have to open a door or move the netting, that time is the same. We'll assume the walk-out time and care time for these birds is 20 minutes per day. You have to go back in the evening to gather eggs and if you have to close a door on them, probably need to go back for that later.

Yes, you can get automatic doors these days, but they are problematic. Aggressive predators love to come at dusk and pick off the last couple of birds not yet in their night quarters. The point is you still have to go out twice a day. We have automatic doors on our remote Eggmobiles on rented land several miles from the farm. We also see piles of feathers from time to time. But in this case, the shut-up time is a lot more expensive than on your homestead where things are close and you can send one of the children out to do the chore. The Eggmobiles we have at our home farm receive three visits per day: morning move and open doors nest boxes, 4 p.m. to gather eggs and close nest boxes, then right after dark to close the doors.

Now think about the totally enclosed birds in a mobile, floorless coop. Going out morning and evening is the same as the other birds. The only difference is that you have to move it, which takes one minute or less. If we say our time is worth $20 an hour, that move minute is worth 33 cents. If our layers are worth $15 apiece, eliminating just one predator attack every 45 days (assuming the predator only takes one bird) pays for the daily move.

Folks reluctant to totally enclose their flock in a mobile coop naively think that predator losses are either not significant or that moving the shelter each day is too much work. It's not. I can't imagine running a flock of less than two or three hundred in a vulnerable open-field situation. The risk is too high; the chickens are too vulnerable.

On our farm, we now run about 5,000 layers in Eggmobiles and Millennium Feathernets. A few years ago when we had an exceptionally bad streak of aerial predator attacks, we ran figures on a totally enclosed system to see how many birds we would need to lose in order to pay for the daily all-enclosed shelter move time. It came out to 30 percent. In other words, when we hit 30 percent predation loss, we'd be money ahead to put them all in enclosed shelters and move them daily across the field. We've not approached that yet, but if the flocks were smaller than 1,000, I think we'd be there in a minute.

I've been conservative with this discussion. The most common occurrence is a predator that wipes out a dozen of the 24 chickens in one attack. Then how many days can you justify moving a shelter every day over letting the birds out to free range? Let's see, 12 X 45 days each is 540 days. That's about a year and a half. That's a lot of protection, don't you think?

And don't think since the chickens are close to your house, they're immune to attack. The problem with homestead flocks is that the small acreage means the birds never move very far. Goodness, sometimes we move our birds half a mile at one time. That really throws the predators into a tailspin--"What's that? Where did all this suddenly come from? I'm scared about this intrusion in my space." All sorts of conversations happen in predator homes when something this big changes in their routine backyard. But a small acreage doesn't allow that level of predator surprise and the inherent off-balancing that it creates.

Lock 'em up, folks. I know they're pretty out there running around. Okay, let them out when you're working in the garden or doing something nearby where you can keep an eye on them. Chickens are the most vulnerable animals you can have on a homestead; protect them.

Chickens love dust baths. Mites are perhaps the most common chicken malady. Mix one part soil, one part sand, and one part wood ashes in a shallow pan for their dust bath. That's a good deterrent to mites. A winter clean-out of their nest boxes and spray-down with dormant oil spray will help keep mites out of the nest boxes. Mites can't live on metal surfaces; they need wood, straw, hay, bedding.

Hygiene requires keeping the floor bedding clean and not capped. It also means giving the birds inviting places to dust bathe, rest, and scratch. Birds like to scratch about as much as cats like to lick themselves. Making sure they have stuff to do keeps them from being bored and gives them lots of exercise.

Cannibalism means some basic need is unmet. It could be nutritional or psychological. This is why I hate seeing birds on wire mesh. If a bird's natural instinct is to spend hours a day scratching, what could be more unnatural than a mesh floor? Battery brooders are horrendous in this regard. Give the bird work to do; she loves it.

Your chickens relish meat scraps. And if you happen to shoot or trap a possum or raccoon, take a knife and open up its belly and toss it in to the chickens. They'll love it. I always get a sadistic joy out of seeing chickens eat up a raccoon carcass. Old timers tell me that the first chore for farm boys long ago was a weekly rabbit, squirrel or vermin to throw in the chicken pen in the winter. They didn't need this extra protein in the summer because they had grasshoppers, maggots, and kitchen scraps. But in the winter the birds routinely became protein deficient. If you happen upon road kill, bring it home to your chickens. They'll peck off every morsel of meat and flesh and then you can throw the bones on your compost pile. All good.

As we close out the laying hen chapter, we need to talk about the bane of egg production: dirty eggs. Our benchmark is 10 percent dirties; if you have to wash more than 10 percent of your eggs, something is wrong with management. Here is our protocol for making sure we have clean eggs.

1. **Oyster shell.** Always keep free choice oyster shell in front of the birds so they can replace the calcium going into their shells. Maintaining shell thickness is the key to clean eggs. Most dirty eggs are not dirty from manure; they're dirty because an egg broke and painted its neighbors with the insides.

2. **A 6 inch lip** on the front of the nest box. Privacy is the issue here; a hen likes to go into the nest box and be as secluded as possible. If you were laying an egg, would you want the world to watch you? Of course not. She can climb in a 6 inch opening, don't worry. That high lip in front means she can nestle down with nothing showing but her head. Discreet nest boxes a foot cubed with a 6 inch front lip guarantee enough privacy.
3. **Nest boxes mounted higher** than eye level. This is probably the most common error I see in my homestead travels. Even most of the sophisticated chicken mobiles for purchase make this mistake. You don't want the chickens to be able to look into a nest box unless they're going in to do their business. You don't want them loitering, looking in nonchalantly, and getting mischievous ideas. Make sure those boxes are above their head height.
4. **Exclusion perches** serve two functions. If the nest boxes are above chicken head height, it means the chicken can't get in from the floor. If she jumps up that high, she'll be flapping her wings and she can't flap her wings and enter that narrow entrance all at the same time. She needs to be able to hop up on something in front of the box, then find the box she wants and step in. This is simply chicken physics. The perch, then, gives her an initial landing place in front of the nest box.

   In addition to her initial landing zone, the perch board serves as a night time exclusion. A simple pivoting hinge allows the perch board to be lowered in the morning and raised at night to keep the birds from being able to sleep in the nest boxes. As you know, chickens leave manure wherever they sleep; if they sleep in the nest boxes, they'll soil them with manure, which will then make the eggs dirty. Your nest boxes should be manure free, always.

If you gather eggs late in the day, like after 4 p.m., you can gather and close the nest boxes at the same time. Hens lay the majority of their eggs before noon.

5. **Use junky hay for nest bedding**. If you use bright straw or nice hay, they'll find seeds, leaves and other goodies to peck. You don't want the chickens enticed to the nest boxes to peck at things. You only want them to come in when they have business to do. Old moldy hay is ideal. A chicken doesn't want anything in that hay and will leave it alone for the most part.

   Now a word about roll away nests. I don't use them. This may sound a bit mystical, but hear me out. What's the first thing a chicken does when she enters a nest? She snuggles down, picks up things with her beak and moves them around a bit as she scrunches into the nest. No matter how big a hurry a hen is in, she always takes the time to sort out her nest a bit. A roll away nest prohibits that kind of basic hen behavior. I think roll away boxes are a cure for a problem people should fix by implementing better design and management protocols.
6. **Install one box per ten hens**. Any fewer and you'll have broken eggs from too many in a nest. Realize that the hens will not use the boxes equally. They choose based on a host of factors, the most important of which is light.
7. **Keep the nests in as dark an area** as possible. The only time a chicken looks for dark places is when she's looking for a nest. Otherwise, chickens love light. Placing the nests toward the north can make a big difference in how much time they spend in them and how they use them. If you're having trouble with the chickens laying in corners or under things, check the orientation of your boxes and make sure they're not open

directly to the morning sun. You might need to modify your windows with some shades or move the boxes farther back into the recesses of the shelter.

If your coop is mobile, you'll notice preferences based on orientation. The hens will use whatever nest boxes are darkest. In our Eggmobile, the birds prefer different banks of nest boxes day to day depending on orientation toward sunlight. The ones that receive the most direct sunbeams don't get used that day.

These are the critical factors. Now, if you have an egg eater or if you develop an egg eater (that's a chicken that goes hunting for a nest box to loiter in just to break an egg and eat it) you can add one more element to this list: install a veil at the top of the boxes, like a curtain. It can be made out of any type of opaque material. Don't worry, the hens will push through the cloth to get into the box. But that will darken down the interior so much that no chicken wants to enter unless she's laying an egg.

Obviously books upon books are available with rations, brooder designs, coop designs, breed explanations and other information about laying hens. I've read many of them and my intent here is to present material I generally find missing, or at least not being done, on homesteads. These tips hopefully will make you enjoy your chickens more than you ever imagined possible.

Chapter 17

Broilers

If laying hens are my favorite (partly because I like eggs so much) then certainly broilers, or meat birds, are a close second. In business-speak, eggs provide cash flow and broilers give quick turnaround.

Goodness, from chick to plate in 8 weeks is pretty impressive; that's about as fast as a radish. Harking back to the previous chapter, you can certainly eat cockerels of dual purpose birds for meat. If you're starving, you can eat any bird for meat, but the return for the effort diminishes a lot as you go to smaller and bonier carcasses.

A good purebred heritage type bird like a Barred Rock, Buff Orpington, White Plymouth Rock or New Hampshire Red should weigh nearly 5 pounds as a mature hen and a cockerel will approach that at about 16 weeks. Of course, cockerels will get bigger if you let them go ahead and mature, but they get tough to process and tough to cook. If you cook long enough and slow enough, even a crow can be edible. Don't discard tough birds; just resolve to cook them long enough and slow enough and you'll have wonderfully edible, tasty meat.

Perhaps one of the homestead's most important pieces of equipment is a good slow cooker or crock pot. Today we can add the Instapot to that arsenal. When we talk about livestock infrastructure we don't generally include kitchen equipment, but in this case I think it's critical. Much of the efficiency on a small holding comes by reducing waste and increasing salvage utility.

If you're up on world wide food waste, you know that the conventional industry has no place for blemishes. I had a friend who worked for one of the big poultry brands and she told me that they had just discarded 30,000 broilers because the farmer hadn't withdrawn feed from them in time. Some feed in their crop and intestines still containing manure meant that the whole lot had to be land filled.

When we process by hand rather than mechanically, we can work around an oops like this and salvage the carcass. With a little more time and care, we can get those guts out without contaminating the carcass. Of course, we like to bring the birds to processing fasting too, just like the industry, but unlike the industry, hand evisceration gives us more wiggle room to work around situations that aren't quite as perfect. This illustrates the inefficiencies and fragility of the allegedly efficient factory food systems.

Another friend who works for the U.S. Civil Service, stationed in Zimbabwe, returned for a visit stateside and told me he had just toured a green bean processing plant. About 4 tons a day came in the front door but only 2 tons a day went out packaged for sale to the European Union. What happened to the other 2 tons? "Oh," he explained, "they were too long, too fat, too short, too crooked--anything that didn't fit the box."

We homesteaders don't throw away misshapen green beans. We snap them and freeze or can them, regardless of size and shape. That is one of our signature efficiencies in the food system and a major contributor to food security. It's also a great object lesson for how small is beautiful at creating a platform to honor the otherwise dishonored. That has some heady spiritual ramifications, no?

Back to broilers. As a homesteader, then, what the industry discards or what has no place in the conventional marketplace can be a significant portion of our household food system. These older, skinnier, slow-growing birds don't look like what's in the store, but they'll eat more grass, be more succulent, and have yellower fat. They can scrounge more for themselves because they're more active.

Remember that exercise makes juicy muscles and lack of exercise makes drier tissue. On the other hand, exercise makes tougher tissue while lack of exercise makes more tender tissue. That's why a thigh is juicier but tougher than the breast meat. The more active a chicken is, then, the more succulent--but tougher--the meat will be. If you want superior taste (which some would argue is an indicator of superior nutrition) you want a more active bird. If you want more tender chicken, you want a less active bird. Age also plays a factor: the older an animal, the tougher the meat.

If you want fried chicken, you'll find the old-fashioned birds a bit of a challenge. The difference in tissue toughness between 8 weeks of age and 16 weeks of age is profound. And of course an old stewing hen (cull layer) is off the charts. Once you've processed a few at different ages, you'll come to appreciate the younger, more tender birds.

I'm belaboring this whole idea of using the cockerels and stewing hens to encourage you to enjoy eating them; don't throw them away just because they aren't supermarket-picture perfect. Enjoy creating value where the industry finds nothing but loss; defy the waste and honor the unconventional. It'll do you good and it'll give you deep satisfaction in your homestead.

With all that said, nothing compares to the efficiency of the fast-growing broiler chicken. The industry standard is the Cornish Cross, but we now have Freedom Rangers (genetics from the French Label Rouge strains), Kosher Kings, Robust Whites and others. These upstarts grow about 10 percent slower than the industry-standard Cornish Cross.

These new competitors in the broiler space claim better livability and taste. Our experience questions those claims; we've tried some of them and keep going back to the Cornish Cross as the best all-round bird for the time and money. But time will tell; now that the industry is having such a problem with woody breast (the breast grows so fast it outruns proper development) perhaps the superiority of the industry standard will lose some allure.

I know I'll irritate some folks with this explanation, but if you want food your kids will eat and food that looks quasi-normal, the Cornish Cross is the broiler of choice. Here's why. A homestead typically is not about making a living; it's about making a life. Your living is derived from someplace besides your homestead. If you wanted to become a full-time farmer, you'd be reading *Your Sucessful Farm Business* or *You Can Farm*, not this book. We homestead for self-reliance and family security. We also homestead to extract ourselves from the industrial machine in the most intimate and foundational elements of life.

As a result, we have to juggle our animal chores around our main income source. The more bullet-proof our models and the more efficient, the less helter-skelter we'll feel trying to get it all done. Let's assume you need 100 broilers for your family's annual consumption (that's roughly 2 per week). You can do that in two batches of 50--probably a good idea--or one batch of 100. For sake of discussion, let's assume two batches of 50.

The slow-growing birds require 16 weeks to get ready to process and the Cornish Cross require 8 weeks. If you spend 15 minutes a day choring this batch of birds, that's a total of 16.8 minutes per bird on the Cornish Cross (15 X 56 days = 840 minutes, divided by 50 birds). The slow-growing birds will take double that time, or half an hour. Now I know that homesteaders typically don't put time value on our chores because it's part of our emotional therapy, our exercise program, and our spiritual fulfillment.

But goodness, the difference between 15 minutes of labor per chicken and 30 minutes of labor per chicken is more than even recreation can bear. I mean, how long do you need to play with chickens each day? By getting through the broiler project sooner, you have time to keep the beans weeded or plant that plum tree. Or get 15 more minutes of sleep for 8 weeks.

The arguments from die-hard slow-growing advocates follow several threads:

1. Cornish Cross chickens can't walk.
2. The slower growers are way more resourceful.
3. Cornish Cross patronizes the evil industry.

Let's look at these one at a time. If your Cornish Cross chickens won't walk, something is grossly wrong with your care. Having raised hundreds of thousands of Cornish Cross in my life, I guarantee you these birds walk. True, they are not as frisky as the slow-growing breeds, but to say they can't walk is grossly incorrect. If they didn't walk, we certainly wouldn't try to move 5,000 of them every day. They have to get up and walk every time we move them; if they didn't, we'd go nuts. No, gentle people, anyone who tells you Cornish Cross can't stand up or can't walk has read one too many exaggerated animal welfare bulletins.

Number two is that the slow growers are more resourceful. In other words, they eat more bugs, forage, and kitchen scraps. In short, they scrounge better. I agree with this, but it only makes a difference if you're giving them more grass, bugs, and scraps. The Cornish Cross certainly eat everything that moves in their daily shelter allotment; they even chase down grasshoppers. Unless you're going to move them more often or give them way more space, the argument doesn't hold. I'll grant that the slower growers eat more forage, and that can have an effect on taste and nutrient profile. I agree, but again, you need to make sure they're on grass that they like in order to leverage that more aggressive trait. As for kitchen scraps, the Cornish Cross enjoy them just fine too. I don't see a big difference there.

Number three is philosophical which means it might be the most important one of all. The independent hatcheries that sell Cornish Cross purchase their eggs from the industry; these hatcheries do not maintain their own Cornish Cross breeding flocks. The breeders are such closely held secret genetics that even the Chinese can't buy them. This argument is similar to the one that says I shouldn't allow Amazon to sell my books because they are the enemy.

Where do you finally go with this? I won't buy gas because BP is bad? I'd like to not pay any taxes because most of what they're spent on is evil and immoral. So do I quit my job, become a hermit in a cave, and kiss society goodbye? While I appreciate the argument, it smacks of sensational self-righteousness in light of all the other interactions we have with society. That said, if you want to pick this as your hill to die on, have at it. Just appreciate that if you have a smart phone, or drive a car, or order something off Amazon, you've just experienced practicality versus altruism.

I'm a huge believer in incrementalism. Wouldn't it be something if we had enough homesteaders to drop factory farming by half? The scrambling and re-organizing required within the industry to adjust to that new market reality would be fun to watch. If over the course of the next decade this book and others like it moved a million households into the homestead domestic livestock world, it would upend everything industrial and that would be a good thing. The critical first step is success on the ground so we homesteaders don't burn out and we stay excited about the next refinements. That is the real challenge, and getting cheap chicks that grow fast and get meat on the table the kids love is the first step. We can make refinements later.

In reality, the Cornish Cross is not natural. It certainly would not have occurred in wild systems. But nearly all of our food comes from centuries of domestication and genetic selection in millions of gardens and peasant barns. I call these Nascar chickens; they need super high octane to perform at their peak. You can't throw out some scratch feed and expect them to thrive.

You need a complete Total Mixed Ration (TMR) formulated by livestock nutritionists if you want great results. Shooting from the hip on these birds like you might with slow growers won't work. We use Fertrell to formulate our rations and they've never let us down, but I'm sure other good outfits are out there. For starters, you can get broiler feed from your local feed store. It will be comprised of Genetically Modified Organism (GMO) grains, but if that's an easy toe in the water, jump in at that level. Don't let perfection keep you from doing good.

Remember that chlorophyll is nature's number one detoxifier. It's like a Roto-rooter through the body. The key is grass; if your chickens eat green material, they'll accomplish 90 percent of what you're after in terms of nutrition, fat profile, health and vitality. Don't worry about the species of forage; worry about its physical form at the time the broilers encounter it. Fortunately, chickens can and will eat virtually anything that's green. As far as I can tell, they have a palatability index even wider than a goat or pig. They'll eat almost anything, including each other.

What a chicken can't do is eat long blades. You can't move broilers through a hayfield and expect them to eat much. All they'll do is tromp it over, poop on it, then get filthy from the manure on the slick blades and stems. The younger the chicken the more finicky it is about what kind of forage is acceptable. We keep our broilers in the brooder for at least two weeks and then put them out on pasture, depending on weather.

I tried putting them out at a week old once, but they were too small to efficiently navigate the grass. By two weeks they're big enough to walk through it. A bird this small needs short blades, preferably only an inch long, and ideally brand new, tender shoots. That means prior to moving the birds onto any given spot, you need to mow so that new, tender shoots predominate rather than old, tough or long blades and stems. This mowing can be done mechanically or with an animal.

The chickens are forgiving in grass length; anywhere from two to even eight inches is fine. Like most things, you won't hit it perfectly every time. But in general, when the grass gets eight inches tall, we bring the cows or sheep in ahead of the shelter to mow the forage. At homestead scale, you can do a leader-follower system with whatever herbivore you have, whether it be a milk cow, some sheep or a couple of calves.

The main point here is that the physical characteristics of the forage at the time when the broilers are on it bear significantly on how much of the forage they consume. If you need to irrigate a bit in a dry time to stimulate some new tender growth, do it. The key to the pastured broiler is the green pasture.

The second aspect of this ingestion volume has to do with freshness. Remember, animals always eat dessert first. They will always eat far more forage if they get a fresh patch every day than if they have a larger area and only get a fresh spot once a week. The most common push back on the pastured model is the time it takes to move the shelter.

Here at Polyface, our benchmark for moving a broiler shelter with 75 birds in it is one minute. Did you get that? One minute. That means one person with our little customized hand dolly can move 4,500 birds in 60 minutes.

If I could list the biggest pitfalls of homestead broiler production, it would be this list:

1. **Shelter too heavy or bulky** or both. Takes too long or more than one person to move, creating inefficiencies and frustration.
2. **Shelter not moved often enough**. Birds are on dirt without forage and soiled from their own excrement.
3. **Grass too long or too short** (nonexistent).
4. **Feeders too small** and the birds run out. Cornish Cross need feed all the time to reach peak performance.
5. **Discomfort**, either too hot or too cold.

6. **Incomplete feed ration**, resulting in poor performance.
7. **Predation for three possible reasons**: shelter too flimsy or open; birds not totally enclosed (running in a netting or free range), imbalance (too many predators per square mile due to lack of hunting and trapping or insufficient ecological biodiversity).
7. **Processing set-up not ready** in time.

Without repeating the material that's in *Pastured Poultry Profits*, remember that these birds grow fast and before you know it, they'll be ready to process. As I mentioned at the front of the book, get your processing ducks in a row before you ever order chicks or you'll never be ready in time.

The best ongoing source of information for pastured broilers is the American Pastured Poultry Producers Association (APPPA). With chat rooms, newsletters and annual gatherings, this organization promotes and informs on all things pastured poultry. It ought to have a million members. As David Schafer of Featherman Pluckers notes, we should have as many backyard chicken plucking machines as we have garden tillers. A backyard flock of chickens should go hand-in-hand with a backyard garden.

Make sure your birds come to processing fasting; withdraw feed but not water at least 18 hours before slaughter. Even 24 hours is fine. The chickens will get hungry, but a little hunger never hurt anyone. A clean digestive system is the key to efficient processing.

For you first-timers riding into this with stories of hatchets and chopping off heads . . . DON'T! Grandma did not always do it the right or the best way. The whole idea is a painless and swift exit, best accomplished with a sharp knife to the jugular vein. Instant brain loss, essentially a Kosher or Halal technique. And no, despite what the animal welfare cultists say, all Kosher and Halal slaughtering is not animal abuse. You don't need electrocution, gassing, or oxygen deprivation to do a humane kill.

Fortunately, small-scale processing equipment is now readily available. Don't put yourself through the agony of hand picking unless you're only doing two. Any more than two, you'll want a low-cost automatic feather picker. Scald water needs to be about 145 degrees F with soap to break surface tension. Either rotate or dunk up and down to get good feather flocculation and better scald water penetration around the feather follicle.

Evisceration is on a bunch of videos and in books. The overarching idea is to proceed systematically. Take it one step at a time, like separate procedures. When you develop mastery on your 1,457th chicken, it'll look more like a flow, but if you appreciate each action as a separate action you'll learn the skill faster.

If the carcass has a wound, cut it off. Blemishes are cosmetic and don't damage the meat. By definition, a bruise cannot happen after death. If a bird has a green leg from gangrenous deterioration from injury, that was a wound that occurred long before processing day. Cut that leg off but the rest of the carcass is fine.

The abnormality that will set you back a bit is when the bird has pneumonia. When birds fade away in a day or two and exhibit blue combs, that's normally pneumonia. You'll see a lot more pneumonia on the shoulders of the season when the birds might get cold. For homesteaders raising only a batch or two, I recommend not pushing the season on either end. In Houston, your best season may be in the winter. In Minneapolis, your best season will be June, July, and August. Don't complicate your situation by pushing the season like we do on a commercial scale. Take the easiest weather time and enjoy it.

If the body cavity is full of yellow gunky liquid, that's pneumonia. The meat is still fine. I wouldn't sell that bird, but again, for personal consumption it's fine. I've lived on rejects all my life. I sell the pretty ones.

Keep your processing surfaces as clean as you can. Keep flushing them with clean water; don't let them build up with blood and guts. Don't be as concerned about the type of surface as cleanliness. A wooden table, piece of metal roofing, or linoleum tablecloth are fine surrogate surfaces if you don't have stainless steel, but keep them wiped clean and flushed.

You can compost the guts but you can also bury them. If you do bury them, I suggest placing a wire mesh on top of the ground when you're done to keep critters from digging into the spot. Just a piece of poultry netting will work fine.

For composting, I prefer wood chips with vertical sides held up by some sort of panel or used pallet to prevent blow-outs of the soupy material. Layering with guts and chips lasagna style works well. I maintain a saucer-shaped top to keep the blood, feathers, and guts from going clear the edge and inviting vermin and odors. Everything but the bones decomposes over a couple of months. We use extra carbon initially because the guts and blood are not very solid. To fully utilize the carbon (chips) we re-use the compost for a second go and then a third cycle. This turns the pile and gets the carbon more fully broken down. While the third cycle still doesn't yield potting soil type compost, it's good stuff and will bring on clover in your fields better than planting seeds.

I like a two-stage cooling system for the finished carcasses. From kill to initial chill should not be more than 15 minutes. We use well water only for an initial chill; the water turns pink and takes the initial body heat away in about 15 minutes. After that, the birds go into ice water. You could use air chilling, which requires turbo air movement and is a bit more sophisticated. You want the internal temperature of the bird to be 40 degrees F within three hours. If you're just doing a couple of birds, you could place them in a pan in the refrigerator, covered, and be just fine.

Never put the birds in a freezer until the internal temperature reaches 40 degrees F. Like all meat, the muscle tissue needs to go through a relaxation time. With herbivores it's usually longer than omnivores, but all meat needs time for the connective tissue to relax and become tender. If you want to cook a bird immediately after processing, be sure to use a long, slow cooking method.

But if you're going to broil it or fry it, you want all the body heat out and 24 hours of relaxation prior to cooking. Many first-timers are disappointed with their chickens because they're so eager to try one they kill it and cook it in the same day: big mistake. Let it sit in the refrigerator over night and cook it the next day. You'll be surprised what a difference the wait makes.

For the same reason, you don't want to freeze the bird too fast. Frozen tissue does not relax, so a fast freeze interferes with the natural relaxation process. Another reason to wait for freezing until body temperature gets below 40 degrees F is that homesteaders too often put 20 or 30 freshly-killed chickens in their back porch chest freezer. These non-commercial freezers aren't made to handle too much ambient-temperature stuff at once. The birds on the inside of the pile may not fully freeze for a week. That's fine if they go in at 40 degrees F. But if they go in at 70 degrees F, they might be two days coming down to 40 degrees F and that's a big problem. Did I say I've been there and done that? You don't want chicken that smells like fish. Buy yourself some insurance and get it down to 40 degrees F quickly before putting it in the freezer.

Because of the large body of information I've already developed around pastured poultry, I didn't want to repeat all the nuances of that system here. But to be fair, you don't have to raise your broilers in a portable pasture shelter. If you have a postage stamp yard, you can use a deep bedding system in a stationary coop. If you opt for that due to insufficient land, plan to supplement them with lawn clippings. The birds will love pecking away at fresh lawn clippings.

In order to keep the grass clippings palatable, you can put some in a plastic bag, press out the air, and tie it off tight. It will ferment in there, making silage. Chickens love silage of all types; they naturally eat fermented material when they scratch into the anaerobic soil layer. Fermented feedstocks are perfectly suited and palatable for poultry. In this way, you can mow the lawn once every two weeks but distribute the clippings in highly nutritious form every day in between mowings.

If your lawn is too small to generate many clippings, see what's in your neighborhood and available for clean-up. I don't have all the answers here, but I want you to think about how you can get these birds something green because that's the signature dietary component. Many years ago when we were getting started and we would process chickens for others (illegally, but we didn't ask and didn't tell) we partnered with a family who raised a set of Cornish Cross broilers in a vacant shed on their property.

We collaborated on our chick order; they took half and we took half. We raised ours on pasture and they raised theirs in the building. We used essentially the same ration. We processed them the same day. All the chickens I've ever raised have been on pasture or some permutation of outside-themed models. We used stainless steel chill vats and never had any fat on them when we did our birds. We'd drain the chill water when we were done and the tanks looked perfectly clean.

But these indoor chickens, bereft of anything green in their life, left a heavy coating of greasy fat on the chill tanks. It was like we'd smeared Crisco on them and we had a terrible time cleaning it all off when we were finished. I'd never seen that before. A couple of years later, we had the same experience with a set of rabbits that Daniel helped a guy process. Daniel's pastured rabbits would squeak when we ran our hands along the wet skinned carcass, kind of like clean hair after a shampoo. The other fellow's rabbit carcasses, never offered green material, felt like Vaseline. They were so slippery we could hardly hold them in our hands. Even though salad ingestion (green forage) may not be the majority of the diet, it packs a powerful punch.

My bottom line: jump in even if your set-up is far from perfect and even if you have tons of unanswered questions. They'll sort out over time and whatever you raise will be light years better than what's in the store. Get started down that path and you'll do better every day. You can count on that.

Chapter 18

Pigs

When does a pig become a hog?

When you can't fill it up.

Pigs are problematic for a homestead because they are by far potentially the most destructive to the land and the stinkiest. When we first started with pigs when I was a kid, we had the proverbial mucked up pig lot, fed them mostly whey and extra milk from the two Guernseys, and carried them lots of weeds from the garden.

And they stunk. When we got rid of the milk cows, we didn't raise pigs for awhile until we began using them for pigaerating--aerating the barn bedding to make compost. Then people liked the pork so much they wanted more of it and we had to go into the hog business, not just the compost turning business.

At that point, I made the Tenderloin Taxi, a 20 ft. X 20 ft. square that I pulled around with the tractor. We had four hogs in it and sometimes they tore up the ground down a foot and sometimes they hardly touched it. That's when I learned that soil moisture is the determining factor in how aggressive they dig.

That kind of gave me my square footage ballpark of about 48 square yards per hog for 5-12 days. We never leave a hog in the same place for more than 12 days. Again, all the infrastructure and specs are in *Polyface Designs*. Here I'm going to mention a few things that aren't in there, and especially scale it down to homestead size. Instead of talking about groups of 50, we'll talk about 2 at a time.

Hogs are extremely social animals and will do far better in pairs than in singles. Try as you might, your human-ness can't substitute for a true pig mate. If you don't need but one, raise two and sell one. Like everything else I've talked about on pasture, movement is the key to everything. If you leave a pig too long in a spot, the area will turn into a moonscape and take forever to recover.

Growing pigs cannot over eat, meaning you can put feed in a self-feeder that can hold them for several days and they won't waste it. They'll eat it as they need it. For a couple of hogs, you can purchase small, simple self-dispensing feeders. The main thing to remember about pigs is that they love to get their noses under things and tip them over. If you use a nipple waterer or a feed trough, you have to fasten it to a sled big enough and heavy enough that they can't tip it over. An easy way to do this is to make a couple of skids with a floor so that when they step up to drink or eat, their feet are on the sled. They can't get their noses under something their feet are already on.

The other thing about pigs is that they are wizards at making mud holes. They will fiddle and fiddle all day with a watering nipple or water trough to either tip it over or waste water and make a wallow. Pigs don't sweat so they are extremely susceptible to sun scald and sun burn. The mud plaster not only cools them, but also protects them from sun damage; the lighter colored the pig, the more susceptible to sun burning.

Up until 150 pounds, pigs should have two strands of electric wire on their fencing. After that weight, only one will work just fine. If you're bringing in new pigs, you can train them by putting them in a secure pen and cutting off a corner with a single strand of electric fence with a spring in it. The spring keeps them from breaking it as they're being trained. If they respect it for 3 days straight, you can take them out to pasture and they'll stay in.

Small pigs have soft teeth and need more protein and concentrated nutrients. As they age, their teeth get stronger, enabling them to eat more coarse things like walnuts and whole shelled corn. The older they get, the more forage they can eat and metabolize as a percentage of their diet. We like to take our pigs on up to 300 pounds or more because only after 200 pounds do they become efficient at grazing.

At this point, we need an interlude to talk about breeds. By now you know I'm fairly color blind regarding breeds. I break them into the exotic forage hogs versus the standard varieties. Kune Kune and Mangalitsas are the most common grazing hogs and they will actually grow without grain but it takes two years. This throws us in a similar tension that we had with the 16-week heritage roosters and the time investment.

A typical hog breed like Yorkshire, Hampshire, Berkshire, Tamworth, Duroc and others will finish out at about 8 months old. If you're buying a weanling pig at two months old, you'll have that pig an additional six months. If it takes 15 minutes to check and feed (chore time) the pigs daily including a weekly paddock shift, that's 2,700 minutes over the 6 months. Since you have two, divide it in half, giving 1,350 minutes per hog which is about 22 hours. We won't put a dollar value on that time because you're not doing this commercially.

Now imagine if you need to keep that hog for 22 months. Same chore minutes per day yields a total of 9,900 minutes. Again, since we're raising two, we'll divide that in half which is 4,950 minutes per hog or 82 hours. Do you really want each of your hogs to require an additional 60 hours versus the more standard variety? Some people would say it doesn't matter and that's fine; I'm just pointing this out because most homesteaders will never do the math before making the decision.

Realize that these exotic forage hogs don't get as large as the standard ones, either. When you take them to the abattoir, the kill fee is standard regardless of size. Your trip there costs the same in time and running whether you take 200 pound hogs or 300 pound hogs. The bottom line is that the smaller weight gives another $100 prejudicial hit at processing.

I'm not arguing anyone toward or away from any hog; I just want you to make your decisions with your eyes wide open. Realize that all decisions have trade-offs, despite what a salesperson, breed association, or urban foodie might tell you.

In this case, if the standard hog eats $500 more worth of purchased feed than the exotic forage hog, that is no small amount and needs to be factored in on the other side of the ledger. Standard hogs will eat roughly 3 pounds of feed per pound of live weight. In general, standard bred hogs in the growing phase will trade off any feed savings from forage with additional calories required for activity. Hogs running around use up a lot more calories than sedentary ones. While some folks don't like to admit it, pastured hogs do not eat less feed concentrate than their factory farmed counterparts. The pasturing is not primarily for reduced feed costs; the pasturing makes a happy animal and a meal worth eating.

The one genetic element that we do consider important is phenotype. The boxy looking extra long hogs favored by the industry do not hold up outdoors. They are fragile to the elements and have no energy. We call them Arnold Schwartzapiggies--their rear quarters stick out way farther than their rib cage and they tend to look like they came in a cardboard box. They put on no fat and walk arthritically. If you go to any FFA or 4-H market hog show, that's all you'll see.

What we prefer is what we call a torpedo phenotype. That's curved down in the front and the back, but straight from shoulders to side to hams. We call it state of the art 1950s genetics. Those hogs will put on some fat and enjoy being active.

This brings us to one of the biggest issues around homestead pigs: to farrow or not to farrow, that is the question. I advise against farrowing for a number of reasons. Remember we want two pigs; we don't want one. Two sows farrowing twice a year, at an average 7 piggies each, will produce 28 piggies. What are you going to do with 28 pigs? That's well-nigh a commercial scale operation.

The single biggest mistake new pig raisers make is farrowing. What came home a year ago as two wonderful bred sows in short order turns into 30 pigs. Few people can handle growth that fast. And then of course you need to breed them. A boar for two sows is extremely expensive. You can take your sows to someone who has a boar. That's fine, but you're going to have to be happy with the way that farmer feeds and treats his pigs. More and more, it's hard to find someone who will open up his breeding unit for unvaccinated, unmedicated sows from a hippie dippy unorthodox homesteader.

Mature pigs need to be limit fed, meaning you can't just put them on a self-feeder and walk away. They will over eat, get way too fat to breed, and then you have to take them to Weight Watchers. That always creates a scene and I highly recommend against it. On the other hand, it might be just the incentive to tip the scale for someone trying to make a commitment. ha! The practical implication of limit feeding is that you have to feed them once or twice a day, every day. Christmas. New Year's. Thanksgiving. On your birthday. When your wife is having a baby. Goodness, if you're going to get that tied down, you may as well milk a cow.

One of the biggest reasons we don't farrow at Polyface is safety for our many visitors. A sow is protective. We don't want to see headlines in the morning paper "Three-year-old eaten by sow at local farm." Not a good thing. If you have any children or visiting children, I'd steer away from farrowing. Piggies are generally easy to find in your region once you start looking. In any given rural area, a certain number of starry-eyed folks, whether for nostalgia or fantasy, start farrowing a few sows. Be their best customer. They'll love you to death for taking some piggies off their hands. And yes, these would be the folks who didn't read this book.

Always bring the piggies home to a secure place. Unless and until you have a pig-proof pen, you aren't ready for pigs. If you have a cattle trailer, you could use that as your secure pen and put your training wire in there. Please, please never turn out an untrained pig into an electric fence lot and expect it to stay in. Invariably that pig will go up to the wire, hopefully slowly, sniff it and bolt through. You'd think the pig would back up if shocked on its nose, but no, half the time the pig bolts forward instead of backward.

Then you have a pig chase marathon on your hands. Have you ever chased a pig? It might look small, but that pig has endurance beyond your wildest imagination. It'll make a fool out of you before you can turn around. That little 40-pound bundle of beauty squares off, eyeing you, and then spurts away lickety-split for the garden. By the time you catch up huffing and puffing, the little beast has rooted up two tomatoes and ripped through the trellised sugar snap peas.

I won't belabor this scenario because I think you've gotten the point: never turn out untrained pigs into an electric fence lot. Train them in a secure place and then it'll all be easy. If you have two pigs, you'll probably want to start with a paddock about 100 square yards. Envision something 10 yards X 10 yards. They'll be in there a week and then move to an adjacent paddock the same size. You'll want enough paddocks to offer about 80 days of rest before coming back to the first one.
That means you want about 12 paddocks.

Of course, you don't need to build them all at once. You can build them as you go. You can take down the vacated ones and re-erect them ahead, like leap frogging. Now a word about moving pigs. Remember that pigs don't see well and their eyes are close to the ground. When you walk among them, all they see are your legs, which to them look like moving traffic cones. Ah, to a pig that looks like a fun obstacle course. You can catch chickens and turkeys and put them where you want. You can shove a cow, sheep or goat. But you will never shove a pig. You need to outsmart a pig, and since a pig is the smartest of the barnyard animals, take time to plan what you're going to do.

Pigs figure out real quickly if you're trying to make them do something they don't want to do. Getting your strategy worked out beforehand is critical because you only have one shot. One of my most poignant teenage memories was trying to load two pigs to go to the abattoir. My older brother and I helped Dad get these two hogs (yes, they were hogs by this time) into a tight 4-gate pen. Our plan was to walk them over to the trailer in this moveable pen. Each side was about 4 feet, so it was fairly tight and all three of us were pretty stout guys.

Do you think we could move them? Not on your life. We were a total of probably 475 pounds with a center of gravity at about 3 1/2 feet off the ground trying to move 600 pounds with a center of gravity about 14 inches off the ground. I've got news for you, folks, that doesn't work. We were trying to get them loaded before my brother and I had to go to school. Dad was going to take them to the abattoir on his way to work.

I kid you not, we were lathered up like a race horse after an hour and could not get those pigs to go to the trailer. As the pigs got madder, they began lifting up the gates with their snouts. Do you know how much a 300 pound pig can nose press? I'd say probably 300 pounds. We ended up just standing on the gates, trying to keep our balance, as the two pigs alternated picking up one gate and then the other. We looked like pistons in a 3-cylinder engine turning on a porcine crankshaft. I wish I was making this up. I still haven't gotten over it.

Dad had to call and postpone the abattoir date. What we did then was use our heads instead. Dad backed the trailer over the electric fence and made a wooden ramp the pigs could walk up. He moved the feeder up inside, put in some fresh straw for bedding, and fed those pigs in there for a week. By the end of the week, we'd go out with milk and some feed and the pigs would practically knock us down trying to beat us up the ramp and into the trailer. When the next week's date with the abattoir rolled around, we fed them, pulled the ramp, closed the gate, and had those suckers right where we wanted them. We learned a lot about pigs that week.

Don't ever get pigs worked up. If you do, you'll lose your religion and still not get them where you want them to go. The most valuable tool for moving pigs is a sort board. You can buy these from farm supply houses or you can make them out of a piece of light plywood. If you go to a market animal show, you'll see all the pig handlers use them in the ring. It's just an opaque panel about 3 feet high and 4 feet long with grip slots in the top so you can move along with the pig and create a mobile barrier. Remember, pigs can't see well so when you move this opaque panel alongside them, they don't realize it's hanging from your hands. They think it's an impenetrable fence.

For sorting and gently moving pigs around, sort boards are life savers. When you're moving them from paddock to paddock, they have a hard time seeing and believing that the electric fence is down where you want them to go. We use homemade wooden gates between paddocks that we move along as we go. The physical opening makes it easy for them to see and go through without having to convince them that the electric wire is down. Pigs are the hardest animals to train to trust you in an electric fence move. A physical gate makes all the difference.

My guess is that with just a couple of pigs, you would be time and money ahead just to wait the pigs out. Open up the fence and put some feed on the other side. If it's hot, throw a bucket of water over there. Pigs love water on a hot day. Pigs don't want to get up from their lounge spot on a hot day; normally you'd have better success by either moving them in the morning or evening, before or after the heat. If they won't wander over and go through fairly soon, just walk away. Leave the feed; leave the water or add some more, and let them be. Within half a day they'll get hungry or curious enough to ease through. Once you've moved them several times they'll get the hang of it and trust you when you open the fence. Any kind of special call for moving (don't use it any other time) will gently train them to know when it's time to move.

If the weather has been wet and they tear up a paddock too much, you can plant some small grain or even corn in the newly vacated paddock. Don't let the paddock be raw too long. One thing you'll notice is that if the soil is wet, bare patches will quickly sprout from the old seed bank. Be sure to change paddocks from year to year and let some of those previous paddocks grow rank and go to seed. That insures an adequate seed bank for the next time you run pigs through the area. In other words, you have a short rotation but then you have a multi-year rotation.

Unless you're using the exotic grazing hogs, your pigs will not do well on just pasture. They'll need a supplemental ration. Our Fertrell ration contains about 20 pounds of diatomaceous earth per ton and that acts as a mild biological parasiticide for worms. The main worm prevention is to keep moving them onto fresh ground.

Pigs are happy to self-feed on root crops. Back when we only had two hogs, we planted a barn corral in Jerusalem artichokes and for nearly a month those hogs dug and ate the tubers. It was one of the easiest hog feeding trials we ever did. I've planted corn and let them hog it down. One of my failed trials was planting corn, Kentucky wonder beans, and zucchini squash. My plan was for them to self-feed on this cornucopia, but right after I planted it turned dry and I never got any germination.

You'll hear folks talk (brag) about the time they did something--like the Jerusalem artichokes--and you'll get all excited about duplicating it. What they won't tell you is how many times they tried something and it didn't work. I planted a bunch of corn in several pig paddocks one time and the wild turkeys swooped in and ate it all. I mean every last kernel. These things always sound great on paper.

One year I planted nearly half an acre in mangles. I figured I'd feed some pigs for a long time by strip grazing them through that mangle bed with electric fence. But alas and alack, my fantasy of 5 pound tubers ran into the harsh reality of weeds and then drought and that whole notion evaporated in a month. We did run the hogs through there but they ate condiments of a few half pound mangles rather than a bountiful provision of tubers. Such is life on the farm; dreams, trials, some wins and some losses. The point is that sometimes just buying the feed is the easiest and most sure-fire way to success.

Pigs certainly need shade. It can be trees or a mobile structure, but they definitely need to be able to get out of the hot sun. If a tree is small and tender, they may kill it by stripping off the bark. Pigs have teeth upper and lower so they can do a lot more damage quicker than a cow. If we have a vulnerable tree we want to protect, we simply make a crude protection with three pallets wired together. We don't attach them to the tree; they stand up by virtue of being triangular corners. By the time the pallets rot, the tree is big enough to withstand the pigs' rubbing. Tree damage is another problem you can greatly reduce by moving the pigs frequently.

I've talked a lot about potential damage, and if that scares you, good, because it's real. On the other hand, if you're observant and watching how they interact with the land, not only will you minimize damage, you can use pigs as a great land enhancement. The disturbance and manure the pigs spread can wake up a piece of ground in desperate need of some ecological exercise. The pigs can aerate, disturb moss and weeds, and bring on some new plants.

If you have a piece of weedy ground, the pig waterer and feeder can be strategically placed to tear things up. We've rejuvenated countless areas infested with brambles and vines by setting the feeder in the middle of a blackberry clump. After a couple of repetitions, the area sprouts beautiful grasses and clovers. Nothing tears up an area like a pig.

Something I've never done but have always been interested in is using pigs to make terraces. If you have a steep bank, you can run electric fence alleys on the contour and over time, the pigs will move dirt toward the downhill wire and flatten out a shelf. At least that's what I've been told. I don't know how wide or how steep you can do this, but we certainly need some folks to play with the concept.

All pigs have four-wheel drive so you needn't worry about them falling over or getting stuck. They can run up a hill you and I can hardly crawl up on our hands and knees.

Pigs are extremely resourceful at finding mast, both soft and hard. In the fall, we designate oak forest areas for fattening and if the acorn crop is good, this can significantly reduce feed costs. Pigs eat all sorts of seeds and nuts. One of the most fascinating things to watch them eat is walnuts. Only older pigs have teeth strong enough to break black walnut shells. You can't imagine how dexterous their tongue and lips are, in conjunction with their teeth. They'll crack that walnut and roll it around, working it until they get the kernel and spit out the shell. They don't care that it takes awhile to do one nut; what's time to a pig?

Having promoted pasture raising so far, now it's time to turn our attention to times when you don't want a pig outside due to wanton destruction. In my opinion, that's during the winter for sure and occasionally during a goopy wet time, like during a hurricane or flood. Anyone who raises pigs should have the ability to get them to a protected place if need be. On a homestead where you might only need to raise a couple hogs a year, you'd want to get the piggies in the spring and butcher them in the fall. In that case, you don't have to worry about the winter.

But if you find yourself with pigs in the winter, I suggest keeping them under roof on deep bedding until seasonal growth starts. Early on, we tried keeping them out all winter and we found it disagreeable for both the soil and us. In the winter, soil is dormant; moving dormant, wet soil around harms its structure and tilth. The pigs literally turn the soil into brick as they move it and then walk on it, move it and walk on it. What creates aeration in the growing season creates concrete in the dormant season.

Caretaking is disagreeable too. You need to keep the spark up on the electric fence when snow falls or it's weighted down by freezing rain. The only thing worse than chasing errant pigs in the summer is chasing errant pigs through snow drifts.

Water freezes and it's all disagreeable. A simple run-in shed with deep bedding is plenty; pigs are stout, much more than chickens. As long as they can get dry and out of the wind, they can handle almost anything.

For two hogs, 40 square feet apiece would be adequate for enclosed shelter. You don't want to let the pigs outside at this time for the same reasons we dealt with on the chickens. Oh, on a warm day they'd love to be out there tearing up a run, but it's an ecological disaster to do that in the winter. Give them plenty of room inside on deep bedding and they'll be just as happy.

Although we don't have a concrete floor in all our indoor winter hog housing, it's certainly a nice touch. Given enough time, big hogs can tear up a foundation and dig holes that will bury a truck. A concrete floor stops them from making divots in the floor of your shed. It's also easy to clean out with a front end loader.

If you do have hogs comfortable inside on deep bedding in the winter, keep them occupied by feeding them some junky hay.

They'll tear it apart, eat some of it, and defecate on the rest. Pigs are fastidious at picking toilet spots and tend to have their rooting area and their toilet area in separate places. They make a nest to sleep in separate from their toilet. The hay gives them something to do and for sure it drops their manure odor tremendously. Again, I haven't seen a study on this, but my hunch is that the hay increases the carbon level of their manure and helps absorb the most odorous elements.

Pigs also love anything fermented. Silage of any kind is a great supplement that can give them some dietary variety and reduce feed costs. You can make poor-boy grass silage in large garbage bags. Stuff wilted grass in, compress it to get as much air out as possible, then seal it. At homestead level, this is the kind of fun pampering you can offer your stock. It's not work if it's fun. And your pigs will love you for it.

While we're on supplements, the number one tonic for pigs is charcoal. All of my pre-1940 swine books mention charcoal as kind of a blanket curative and preventative. Of course, that was back in the day when charcoal was much more of a commodity than it is today. On our farm, we have an outdoor wood burning water stove for our domestic heat and when we shovel out the ashes, we sift them through a half-inch screen to separate the charcoal chunks. We feed those to the pigs and they eat it like candy. A couple of pigs will eat a pound a day.

We've had pigs that looked like death warmed over, let them eat some charcoal, and in a week they're fair show quality. It's almost like a resurrection. If you have a gimpy pig, before calling the vet or getting too concerned, feed it some charcoal. Of all the suggestions I've made over the years, this one has probably received more positive testimonial mail than any. "It's a miracle!" is the most common description.

For those of you who don't think electric fence will work with pigs and are bound and determined to use a physical fence to contain them, please rethink your position. Pigs are strong and unlike other animals, they can dig. A pig-tight physical fence needs to be a fortress and held firm to the ground. Even then, they might dig under it. Of course, a physical fence is not portable, so generally that locks you into a non-rotated dirt lot, which is everything wrong and nothing right. In our experience, pigs are more respectful of electric fence than any of the other animals, probably because they are smart and don't cotton to self-inflicted pain.

Finally, a word about loading. Anyone who has ever had pigs has loading stories. Pigs don't like to get off the ground. They don't like walking on ramps suspended in the air, where they can feel and hear the hollow clump-clump of their hooves. They don't mind jumping up, but they sure don't like jumping down.

A low-boy trailer is usually low enough for them to jump into, but if you put a bale of hay in the crack between the trailer edge and the ground, they'll go in a lot better. To them, that bale of hay looks a lot less intimidating than the steel edge of the trailer and the air space beneath it. Use your sort boards to gently maneuver the pigs to the trailer. Remember if they turn around and oppose you, you're doing something wrong. Just stop, think, and try to see things from the pigs' perspective. Treats are always helpful--a paddy of nice hay, some freshly pulled weeds or grass, a scoop of grain. Pigs are naturally curious. If you can keep from getting them riled up, and give them time, they'll follow their noses toward anything interesting.

If they think you have bad ideas in your head, they'll start thinking, and when a pig thinks, it's not a good day. This may sound over the top, but while you're loading them, talk sweetly to them and tell them you love them and they're good pigs. Beam nice thoughts to them, not thoughts about "I can't wait to eat your bacon." I think animals have a sixth sense about them that can tell what we're thinking, a kind of telepathic radar.

Obviously it makes a big difference on their final day how

you've interacted with them on all their previous days. If you've handled them a lot and have built up a high level of trust, they'll be much more amenable to what you want them to do. If for no other reason than this, it's valuable to move them frequently all through their lives so they get used to your presence and the benefit of going along with your plans. Of all your possible homestead animals, the pig is the one that demands the highest level of Animal Psychology 101.

If you're up on what you can use from a pig, you know that about the only thing that's not useful is its mind. Everything else can be eaten or used in some way. And who doesn't like bacon or sausage with their morning eggs? Pigs are a wonderful homestead addition.

Chapter 19

Sickness

All of my books are sparse on sickness, and this one will be too. I hope that's not disappointing. I hope it inspires you to realize that you can raise a lot of animals with insigificant sickness. I'm a novice about sickness because we don't have much, even though we have thousands and thousands of animals.

By now you should know that here at Polyface we do everything possible to create a healthy habitat. We desperately try not to cheat and in our experience, every single sickness we've had was our fault. We cheated something.

To be sure, sometimes it's selecting the wrong genetics. Weak or incompatible genetics are sometimes honest mistakes. We've gotten weak batches of chicks from a hatchery, for example. Early on, we purchased chicks from several different hatcheries and finally settled on the one that consistently delivered the hardiest birds.

One of the hatcheries that lost out in that selection process had good chicks, but for some reason a glitch in the postal service always kept them an extra day in the mail. That extra day was life changing for the chicks, in a negative way.

Industrial outfits routinely ask me about our bio-security protocols. From "No Trespassing" signs to shoe disinfectants and sign-in, sign-out stations, conventional industrial farms exercise a host of bio-security measures. We've never indulged in that here; my position is that our animals have healthy immune systems. Many farmers view that as a cavalier, presumptive, almost smug attitude: negligent at worst, naive at best.

Perhaps the funniest bio-security experience I ever had was on Kangaroo Island off South Australia. I did a two-day masterclass and one afternoon the group went on a field trip to tour a local farm that was doing controlled grazing and had some chickens. About 50 of us arrived at the property. The region's head of agriculture bio-security was waiting for us.

He had a basin of sheep dip there for us to all step in before we went through the gate onto the property. Imagine this. We're all gathered around, listening to his lecture on bio-security. The basin of disinfectant is in the middle of the circle, like some sort of altar. He gets done and we one by one dutifully step in the disinfectant on our way through the gate.

As we re-assemble inside the gate, we look up to see a whole flock of kookaburra birds fly over and land, some in the field and others in some nearby trees. They set up a terrible squawking as about 20 kangaroos just 50 yards from us jumped over the fence (we went through the gate) and hopped through the field a little ways before stopping and looking back at us.

God help me, I couldn't help myself. I looked at the government bio-security expert, pointed at the birds and the kangaroos, and asked "where's their bio-security?" Everyone laughed except the bureaucrat. He took it as a direct assault, which is exactly what I intended, and promptly packed up his poison altar and left. Forgive me. The whole thing was surreal, like it was planned by someone. This fixation on sterility is nuts. Nature brims with life; we need to be more worried about sterilizing the planet with pollution and chemicals than we do tracking a feather or piece of poop onto someone's property.

The industry considers wild ducks like mallards a hazard. Deer are a hazard. Indigo buntings are hazards. Anyone who considers wildlife a bio-hazard is, himself, a bigger bio-hazard. We can't sterilize and chemicalize our way to health. We need to embrace life's abundance and diversity. The right thing at that farm would have been for us all to have brought a bit of soil from wherever we spent the night to innoculate (baptize?) the farm we visited with preciously needed different life.

For the record and in full transparency, let me tell you every economically significant disease outbreak we've had during our 60 years of operation here at Polyface. Buckle up; here we go on a wild ride.

1. Marek's disease in poultry. Business was booming and we needed a lot more eggs so we ordered a larger-than-normal batch of pullet chicks in the spring. We thought we ordered them in plenty of time to have them field-ready when spring warmed. Timing was crucial because we had to vacate the brooder before the first batch of broiler chicks came.

Heritage breed pullets are pretty tough and can take more weather extremes than the race car broilers. The pullets arrived and all was well but as the day approached when we needed to get them out and prepare for the broiler chicks, the weather turned cold and wet. Cold and wet. Cold and wet. It lingered and lingered. The pullets kept growing, of course, and got crowded in the brooder.

With one day to spare and the weather unrelenting, we had no choice but to get these overcrowded and stressed birds to the field. They started dying. Lots of them. It was Marek's disease, which commercial flocks have been vaccinating for since the early 1900s. My go-to diagnostic handbook is the *Merck Veterinary Manual*. Every livestock owner should have that book. It's written for veterinarians and has a lot of big words, but you can make out the gist of it.

Of course, at the end of each diagnosis and explanation, if a pharmaceutical can fix it, you're told which one. Guess what; Merck sells most of them. But don't discard the wealth of veterinary information in the manual. Merck didn't write the medical stuff. The symptoms perfectly matched what we had in the field, and guess what it said caused Marek's? Stressed, unsanitary conditions.

Our fault. We lost about half that flock. Realize that means half the flock did not die. Often Marek's wipes out the entire flock; 100 percent mortality. We only suffered about 50 percent; as soon as the weather finally changed and we got some warm sunny days, those birds perked up and although they never laid as well or were as strong as normal, they matured and did okay. For the state they were in, they made a remarkable recovery. As the country song says, they were in pretty good shape for the shape they were in.

Marek's stays in the soil for a long time and it was devastating enough that we didn't want to risk getting it again. Fortunately, like many of these opportunistic diseases, it's fairly weak and the virility in a specific location drops fast when things return to normal. We had our pullet chicks vaccinated for the next three years as a precaution and then quit and we've never vaccinated since. That was probably 30 years ago.

2. Curley toe in broilers. I had just written *Pastured Poultry Profits* and things were cooking along well when we suddenly began suffering huge losses from a paralysis in the broilers. But it wasn't just paralysis; it started with the toes curling up, prohibiting the chick from being able to walk. It hit them at about 7 days old, while they were in the brooder.

It stopped shortly after moving them to the field. But it was deadly and we went back to our *Merck Veterinary Manual* for help. Can you believe the index has a listing called "Curled toe paralysis (poultry), pg. 1434?" Yes, it does. I'm telling you, folks, this is the go-to diagnosis book for me.

Do you know what causes curly toe paralysis in poultry? Riboflavin deficiency. But it's more than that. In these fast-growing chicks, the nerve sheath can't keep up with the structural growth of the bird. This internal rub, or friction, creates a staph infection. While I was still in panic mode, I contacted the state vet lab in Harrisonburg to inquire about treatments. They gave me a sheet of paper with about 20 antibiotics on it, but all were scratched out except about 4. The reason? These antibiotics had been used so much in the industry that the staph had already built up resistance, rendering these antibiotics useless.

I'm sure you've heard of super bugs, MRSA and cDiff in hospitals--a host of extra-potent diseases that exist thanks to the over-use of antibiotics in farm animals. I didn't want to use antibiotics, so in a stroke of genius, I decided to find out if there was a natural source high in riboflavin. Also on my bookshelf is the original *Let's Eat Right to Keep Fit* by early healthy eating maven Adelle Davis. She was the food sequel to Robert Rodale's organic gardening pulpit.

Davis charts numerous foods and their relative nutrient value. I found riboflavin and looked at its food sources: leafy greens and organ meats. The lights went on in my head: that explained why the problem stopped a few days after the birds went on pasture. They ate leafy greens and rectified the problem.

But I needed a way to stop it, period. I went to the freezer, pulled out some packages of frozen beef liver, and took them to the chicks in the brooder. To say they attacked it would be the understatement of the year. As it thawed, they pecked off pieces and within an hour had consumed the whole package. I separated the pitiful paralyzed birds and put them in a hospital section to make sure they could get the liver. Within a couple of days not only did the paralysis stop, even the ones that were completely paralyzed stood up and walked.

I believe in miracles, but this was not a miracle. It was the manifestation of a beautiful system God created to keep things healthy. We violated that balance and when we restored it, things returned to health. The most interesting part of this story to me is that not only did the liver and grass stop the paralysis from progressing, it even cured the sick birds in the face of staph infection. That's a powerful testimony to the strength of good nutrition. Fortunately, animals don't have all the psychological maladies that possess humans; their issues are a bit more simple to solve with basic nutrition. With humans, you have to work through emotional stress and all sorts of mental and spiritual stuff in addition to nutrition.

Again, this problem was our fault. We did not give them the nutrition they needed. True, this would never happen in old-fashioned genetics; it only happened in these Nascar rocket-fuel birds. But nevertheless, we did not give them what they needed nutritionally and they suffered for it.

3. Blackleg in calves. We had rented only our second farm property and it was overgrown with brambles--wild blackberries. I mean clumps a couple acres in size and canes 12 feet tall. These impenetrable thickets provided no pasture for the cattle and we wanted to mow them down to start the transition to grasses. The landowner wouldn't let us do that because he wanted the habitat for birds and rabbits and wanted to pick the blackberries when they ripened. I refrained from asking him how many acres he thought he could pick.

What to do? We decided we'd do a work around and use the animals to stomp out the blackberries by putting the salt box in the clumps. We took over a group of about 100 stockers and by the end of the first week, one died. A nice one. These were all robust, healthy, fast growing animals. Finding a dead calf is highly unusual. We brought it home to the compost pile and when I went to move them the next day, another one was dead. Mind you, not a single animal showed any stress of any kind. The day before, except for the dead one, they all moved eagerly and began grazing.

Normally an animal doesn't go from healthy appearance one day to dead the next. Now we'd lost two and were concerned but still thought we'd wait and see if this was just some sort of fluke. It wasn't. The next day I went over and two were dead. We realized we were heading toward a catastrophe. This all developed in the first few days of our first time on this farm. What could possibly be causing such an acute reaction? We looked around for wilted cherry leaves or yew, the most common toxic things in our area, but couldn't find any.

In desperation, I called the vet and he came over, took out a scalpel, and cut open one of the dead calves' back legs. The muscle was black. He looked at me with that vet determined look and announced: "blackleg." I asked him what I should do and he said to vaccinate. We never vaccinate anything, but I also didn't want to lose a herd of calves. I drove over to the clinic, picked up the vaccine, and the next morning we vaccinated the herd. We did lose one more that day, but that was it. The vaccine stopped it cold.

This didn't sit well with me because in general, I think vaccines are crutches--or curses, depending on who you read--and shouldn't need to be used if things are in the right natural balance. I came home and started going through my old cattle books. I collect old farm books for a hobby. I looked up blackleg in each one, and each one gave the same information: ubiquitous protozoa, lives in the soil, can't be eliminated, only cure is vaccination. The mortality protocol was not composting, either (we compost everything, including aggressive bureaucrats). I've instructed my family that when I die, "just throw me in the compost pile."

The way to handle mortalities by the book was to either incinerate or bury, cover with lime, stay away from the area for a million years--you know the drill. Goodness, here we were composting it and then spreading the compost on our fields. What could be more disease inviting?

As I looked through each book, reading the refrain, I became more and more discouraged. Surely somebody would have something different to say or a tidbit to add to the discussion. I finally got to my last book--no kidding, I am not making this up--and read the same stuff but wait! A sentence at the end of the discussion caught my attention: "enters the body through anaerobic puncture wounds, normally caused by brambles." Aha! There was the eureka moment.

Under any normal pasturing conditions, these animals would have never gotten blackleg. But because we'd turned them into pin cushions, the protozoa normally kept at bay had freeway on ramps into these poor calves. Another interesting piece of the story is that all the books agreed that extremely healthy, fast-gaining animals were more susceptible because high metabolism also speeds up the infection. This confirmed my observation that it seemed like the nicest calves were the ones getting infected. Using those calves to stomp out the brambles turned them into pin cushions of anaerobic puncture wounds.

At any rate, I had my answer, went to the landlord with this information, and he gave his consent to mow down the brambles. We did and haven't had an incident since nor have we vaccinated. Yet all our neighbors vaccinate with religious fervor every year. Again, the problem was our fault completely.

4. Turkey coryza. One of our former apprentices had graduated to a subcontractor position on one of our leased farms and belatedly told us he was having trouble with his turkeys. We went over and sure enough, they were dying like flies and looked horrible.

Their enlarged head, eyes, and nose indicated acute infection. Their eyes swelled shut and then they died. The problem? Acute lack of sanitation. The water tub hadn't been cleaned for days; it was half full of turkey poop and shoved up against the electric netting. It was probably carrying an electric charge, which made the turkeys reluctant to drink. They weren't being moved on schedule and the paddock inside the netting was filthy.

To say that Daniel and I were outraged would be the understatement of the year. We brought the surviving turkeys home (big mistake) hoping with some extra care we could save them. We put them in a hoop house on clean bedding where we could make them comfortable with the perfect temperature. The disease did ease off, but we still lost most of those turkeys.

Remember I said bringing them home was a big mistake? We found out just how big it was the next winter when the same thing affected the laying hens in that house. We lost perhaps a quarter of those hens and thought we were over it. But no, sure enough the next winter we had significant losses as well in that hoop house. Not the other four adjacent hoop houses, mind you, but only in that one.

The second winter's infection rate was not as bad as the first winter, but still significant. Finally the third winter things were okay again.

I hope you're beginning to see a pattern here. Our poor management creates some sort of breach that opportunistic bugs exploit. It's pretty simple really. Okay, let's keep going. I hope you're enjoying this; I'm going to have to get a glass of milk and eat a cookie when I get done with these horrendous memories.

5. Rabbit diarrhea. When Daniel began his rabbit enterprise at 8 years old, we decided we wanted the rabbits to be pastured or at least forage based. At the time, we didn't know how we would pasture them and realized we might have to take the salad bar to them by cutting it and carrying it in, a method known as cut and carry.

Rabbits proliferate fairly well if you don't make too many mistakes, and his did their rabbit thing. Pretty soon he had a dozen does kindling about 7 bunnies three times a year. He would bring in green material for them to eat and for the first 5 years, experienced 50 percent mortality. Half of all the rabbits died. Fortunately, he did not depend on this for a living at the time and he had a beneficent banker so we put our heads down and stubbornly kept at it.

For background, realize that every rabbit book at the time admonished to never feed rabbits grass because they would get diarrhea and die. We shrugged off that counsel because we reasoned that in nature, rabbits normally eat grass. Suddenly, sure enough, in year 6 he quit losing rabbits to diarrhea. It was like the valve shut off.

What was the explanation? Genetics. Look, if you have 50 animals and one is sick, that animal indicates a weakness. Certainly you can compensate for the weakness by giving it a vaccine, a supplement, a pharmaceutical, but all these treatments are temporary props for inferior individuals. Because he didn't reach for the drugs, he knew which rabbits were weak and kept selecting the survivors. We call this survivor genetics. It sounds harsh, to let function drive form, to let the weak ones reveal themselves, but in the end, that is the best way to find out which ones thrive in your system and which ones don't.

In this case, the problem was less our management and more incompatible genetics. But that was still created by an improper selection process going back who knows how many generations before Daniel got his initial trio (two does and a buck). Perhaps it wasn't acutely our fault, but it was human error that set up the weakness by never feeding forage to rabbits over generations. After generations of nothing but hot commercial feed, the rabbits couldn't tolerate their most natural diet. Isn't that amazing?

6. Sheep parasites. Long time ago we decided to add sheep to our repertoire and purchased a small flock of Katahdin ewes. I was committed to not doping them up with the conventional wormers every month. A monthly worming treatment is the normal regimen in areas like ours that are more temperate than dry.

This story is almost identical to Daniel's rabbit story above. We averaged about 50 percent mortality. Remember, that means 50 percent lived. In a flock of 100, with 50 looking fantastic and 50 dying, all in the same flock, getting the same treatment, and all the same age, what does that tell you? It tells me that some of them don't fit our system and some of them do. The sooner we find the performers, the better for all of us.

We didn't keep the sheep long enough to work through these weaknesses. When our grandson re-introduced sheep a few years ago, however, he got them from more compatible genetic stock (the owner did not use wormers) and he's going on 5 years without worming once and without losing a single one to parasites. Granted, it's a small flock, but he's now up to 50 and going strong. It's quite remarkable.

7. Turkey pneumonia. One year we flubbed our brooding capacity and needed to move a batch of turkey poults outside long before they were old enough. We cleaned up a bay of the barn, put poultry netting around it, made some protective walls out of hay bales, and put in the turkeys.

As often happens in these kinds of thrown-together situations, nature came in with a week of cold, blustery, rainy weather. The poor turkeys didn't have a chance. Poults are extremely fragile and we simply did not have any better place to put them. It was a batch of 400 and we lost every single one of them EXCEPT ONE!

If we had been better positioned, we should have kept that one and started a new high immunity line. What enabled that one out of 400 to not only live, but thrive? He never got sick or puny. Pneumonia went through that bunch of turkeys and wiped them out, but one refused to go down. What a great turkey.

But again, we created the problem.

8. Salty feed. We had a routine batch of chicks in the brooder and suddenly they began drinking water, jumping in the waterers, getting wet, then suffocating as they piled up trying to get warm. It was completely bizarre and I couldn't imagine what the problem was.

I called the owner of the feed company that mixed our feed at the time and he came out, took a look, and agreed that something was strange but couldn't put his finger on it. When he went back to the mill he started asking around and discovered that the bulk truck that had delivered our feed had taken a mineral mix that is primarily salt to a large dairy prior to coming to us with our poultry feed.

The protocol required cleaning the truck thoroughly, including the auger, before putting any other feed in it. That didn't happen. As a result, when he came out to our farm and augered off our feed, a couple hundred pounds of the salt-based mineral mix came out first. It wasn't pure enough to notice because some of our feed was mixed in it, but it was way more salty than safe. The chicks ate the salty feed, went crazy with thirst, piled up in the water, then got wet, then piled up to stay warm. It all made perfect sense as soon as we put the pieces together.

This wasn't a disease, obviously, but it shows the kinds of things that can happen when you're dealing with humans and machines. This is not a perfect world. But again, we created the problem.

Whew! Does all this make your head spin? It does mine. Where's that cookie? The upshot of these stories is obvious: every single sickness we've had was our fault. This is important to understand because it makes sickness a learning experience instead of a victim experience. If you think the sickness fairies are sprinkling sickness from a random shaker in the sky, you'll never learn anything about how to build your animals' immune systems.

But if every time an animal gets sick you look in the mirror and ask "what did I do, or what can I learn from this?" you'll do better, whatever better means. I've belabored this issue a bit more perhaps than is comfortable because the fewer the animals you have, the more special your relationship is with each one. I know our cows only as ear tag numbers. But if you only have five, you might name them. You'll know who the bossy one is and who the timid one is. Knowing these characteristics elevates them to a higher emotional equity level. I get that.

But the principles and lessons illustrated here are just as applicable to a small scale operation. We can do a lot ourselves. We can learn a lot ourselves. We can all do better, all the time. If we put the emphasis there rather than trying to save everything and call the vet over every issue, we'll gradually cultivate hardier animals that are ultimately happier. And that makes us happier.

Probably the most proactive thing you can do for health, besides sanitation, genetics, and comfort, is proper minerals. Due to chemical fertilization and intensive production, feedstock minerals are far below what they were decades ago. Worn down soils don't offer the oomph that rested and rejuvenated soils do.

On our farm, the holy grail of minerals is Icelandic kelp. Sea water mineral ratios are nearly identical to healthy mammalian blood. Kelp uptakes these minerals in that ratio and offers them in chelated form. All of our animals get kelp. It's expensive, but it's a lot cheaper than vet bills and sickness. Investing in minerals is probably the single most important cash expense you can spend to maintain herd and flock health.

Just because the local feed store has livestock minerals doesn't mean it's the right stuff; it seldom is. Go for the outfits who advertise in the publications serving agriculture's lunatic fringe like *Stockman Grass Farmer* or *Acres USA*. Minerals certainly prove the axiom that you get what you pay for. Jerry Brunetti, founder of Agri-Dynamics, mesmerized audiences in his presentations with electron microscope pictures of mineral-deprived versus mineral-satiated gastrointestinal tracts. Without proper mineralization, the body can't uptake nutrients, no matter how good they are.

An animal being picked on needs protection. Fortunately, at small scale, that's seldom a problem. Offer plenty of room at feeders and plenty of room in shelters and you shouldn't have to worry about animals getting rooted back.

The fastest path to health is ruthless culling, assuming we're doing all the right hygiene, dietary, and stress stuff. The sooner you put wheels under a problem animal, the sooner you can put your attention on other things. If an animal doesn't pay its way, it's a liability and will gradually take some of the enthusiasm off your homestead dreams. That's too big a risk to take. Shedding a tear is fine; I've certainly shed my share.

I had a calf once born in a blizzard. I brought it into the barn but it was too weak to nurse. I milked the cow and stomach tubed the calf. The next day he stood on his own. I thought he was okay but a couple days later, he began limping. I checked him over and found his legs cold. Frostbite. We didn't know the extent of the damage so I kept him in the barn and covered him in hay. We fooled with that calf and fooled with that calf for a month until we realized all four legs were frostbitten and would fall off below the knee. I put him down. I'd invested hours in trying to keep the little fella alive, all for naught.

Probably my most tearful parting was one of our milk cows. That's one of the most intimate relationships on a farm. A hand milked cow becomes almost a member of the family. Her name was Whitey because she was half Angus and half Holstein--pretty much black with a solid white head. Somehow she ate a piece of wire and died of hardware disease. I cried over that because of all the milk cows we ever had, she was my favorite. We could have called the vet and spent $1,000 putting a magnet in her and trying to extract it. But we didn't have $1,000 to spend on heroic efforts. She wasn't worth $1,000. Was she special? Yes. Was I sad? Terribly. But she's not grandma. Ultimately she's worth what the market says she's worth and that's how you need to look at these things.

C-sections are never worth it. Normally you lose both cow and calf and you're out $400 for the surgery. We can't fix everything we'd like to. Part of the thrill of homestead livestock is knowing what a good animal is. How do we know what a good animal is? By dealing with difficult ones. We dry our tears and we move on, having lived and loved and nurtured to the best of our ability, gaining a new appreciation for life, the luxury of health, the fragility of relationships. Everything worth knowing you can learn around animals on a homestead.

When do I use a vet? A healthy cow unable to birth deserves veterinary assistance. Fortunately, I know lots of tricks and can do almost everything myself. The best trick I know is the 180-degree turn. The cow's pelvis is shaped like an oval top to bottom. The calf comes out backbone up and belly down. Its widest spot is its hip bones, which can hang up in the corners of the cow's pelvic oval--the north end. By turning the calf sideways and letting those hips come out north and south, you can slip a calf out by hand that you can't get out with a winch.

Always pull calves or lambs down as you pull them out. Don't pull them straight out; remember that naturally they dangle as they come out. Keep that angle of descent the same when you're pulling. Sometimes you'll end up under the cow by the time you're done. Have obstetric equipment handy.

The more valuable the animal, the more you can spend on remedial care. A chicken never pays for a vet call. Nor does a rabbit. A cow may or may not. A sheep or goat, virtually never.

This isn't about disliking vets; we have a vet outfit near us that's fantastic and I'm glad they're here. But if everyone used them as much as we did, the practice would be a quarter the size that it is. When cows drink out of ponds and sheep receive pelletized feed and pigs live in squalor, vets stay much busier.

I hope these stories and the overall philosophy helps to answer the question about when to intervene and more importantly, how to minimize sickness risk. If you want to fixate on something, fixate on providing everything necessary for a vibrant immune system. If we fixate on possible diseases, we squander precious money and time on the negative instead of investing in the positive.

The bottom line is that nearly all the time, we cause our problems because we haven't done something right. Let's be diligent about meeting all the needs of our animals and selecting the ones that thrive; if we do that, nearly all the problems will be marginal.

Chapter 20

Sorting and Herding

All domestic livestock are naturally prey animals. That means their eyes are set wide, on either side of the head, which greatly increases their peripheral vision.

Predators like humans, cats, and dogs have eyes set more in front, closer together, to better focus on a target. A cow can see in a 330 degree arc; only a 30 degree swath, directly behind her, is outside her eyesight. Our human arc is no more than 180 degrees by comparison.

For brevity and simplicity, I'm going to talk about cows for now, but these general ideas apply to all domestic livestock. Cows are probably the most predictable in their response to the principles I'll describe here, but all of them exhibit similar traits.

All cows have a flight zone. It varies from individual to individual. If you move up to a cow, unless she's a pet, you'll eventually penetrate her flight zone and she'll move. When you look at your cow, imagine an aura surrounding her, like a ghost halo, that defines how close she'll let you get to her. If you penetrate the flight zone, she'll move far enough away to feel safe.

Because cows live in the moment and are hyper-aware of their surroundings--literally their surroundings--even the slightest movement triggers a response. A turn of the head, an arm movement, simply shifting your body. Let's start in the field and then we'll come into close quarters at the corral.

In the field, this flight zone is different from individual to individual. A skittish cow, for example, may have a 20 foot flight zone. In other words, if you get within 20 feet of her she wants to step away. Another more docile one may have only a 10 foot flight zone. The big no-no is getting behind her where she can't see you. That stresses the cow because she can't see you. Never, ever, ever walk directly behind a cow.

The flight zone isn't round; it's an oval, shaped like the cow. Imagine the cow standing broadside to you. If you walk straight at her, perpendicular, toward the front part of her shoulder, she'll back up. Her response to flight zone penetration has everything to do with the angle of threat. If the angle comes from the front end, she'll back up. If the angle comes from behind her shoulder, she'll move forward. This nuance is easier to detect in small flight zone animals. If you have a wild and crazy cow on your hands with a 100 foot flight zone (yes, it happens) this shoulder nuance is not as easy to define.

In the field, where a cow has much more room to maneuver, the response triggers are more sluggish than in tight quarters in a corral. Out in the field, you both have more wiggle room. If she whirls around to look at you, it means you either stepped into her invisible zone or you've applied too much pressure.

Herding is an ongoing dance between pressure and release. You enter the flight zone to create movement response due to pressure, but then as soon as the animal moves in the direction you want it to go, you need to release the pressure. Otherwise it feels chased. This pressure and release dynamic is the foundation of all herding. Once momentum toward your objective occurs, maintaining pressure makes the cow hurry up, turn around, or do something you don't want.

As Bud Williams used to say, whenever something isn't going right, stop and slow down. The cows don't have an appointment book. They don't know that the haul trailer is coming at 10:30. They missed your text message. All they know is you're out there chasing them around and they want to get away.

Impatience working cows will never help the situation. It'll compound anything that's going wrong. I know. Guilty as charged.

I'm quite aware that many homesteaders with extremely small herds will have such tame animals they'll follow you anywhere. But if you buy one, you may need these skills before it turns into a dog. Working animals from the flank, out where they can see you, will enable you to apply pressure and ease off in a controlled choreography. Getting up around their heads or in behind them might create a choreography, but it won't be as pretty.

Because every cow has a different flight zone and different personality, when working with more than one you have to figure out the lead animal. By definition, herding animals are followers. Movement draws movement. When one animal makes a move, the rest follow. Their instinct is the safety of the herd. That's why when one animal squirts out of the herd, it usually doesn't run very far before it stops, turns around, and starts wanting to rejoin the herd.

Chickens understand this as well. Here at Polyface we've been vilified by militant animal welfarists who call our mobile broiler shelters jails. They mistakenly think these little chickens yearn to be free, to be outside the shelter. Anyone who knows anything about chickens knows that as soon as one squirts out, it starts circling the shelter, trying to figure out how to get back in. Even rogue chickens in the Feathernet who get out spend most of their time running around the netting, trying to figure out how to get back in with their friends. Even as small as a chicken brain is, flock safety wins over lone ranging.

If you make a mistake and an animal squirts out, don't chase it. Just relax, step back, stop, and let it rejoin the herd. Almost never will one cow head for the hills and keep running away from the herd. Any cow that does that is in a mental breakdown. That behavior just ain't natural.

As you can imagine, the animal that responds first to your movements is the one with the widest flight zone. That animal normally sets the tempo and temperament of the entire herd. That's the animal you play on. Don't worry about the stragglers; they'll come on. Concentrate your attention on that lead animal; that's the one that determines how the herd acts and where it goes.

Often when people herd cows they get all frustrated about the laggards. Don't worry; they'll catch up eventually. As long as you keep the lead cow heading right, the rest will follow. Once you get in the corral, response time quickens. Since the close fences prohibit the cows from being able to move as far away from you as they might like, you as a handler need to be even more cognizant of their responses. And you need to slow down and back away.

Whereas in the field you can usually take the pressure off by simply stopping, in the corral you will probably need to step away. As soon as they see you stepping away and giving them more space, they'll settle.

When you're sorting, a slight movement--not even a step--can start or stop the cows. If you want the full skinny on all this, the book *Moving 'Em* by Burt Smith is the definitive manual. If your herd is extremely small, I encourage you to train your animals to follow a treat bucket. From a handling perspective, if your animals will follow you, finesse and infrastructure can be less.

A little money spent on molasses or some sweet feed can make your herd follow you right onto a trailer. A cow will jump a cliff for two handfuls of sweet feed. That's not grain feeding; it's a treat that can pay big dividends on moving or loading day. Sometimes the treat can be a handful of really good hay. Anything different can work. The main idea is to use the same bucket, the same call, the same rattle (like some sweet feed you shake in a bucket), and ideally the same treat every time. When the animals get used to it, they'll come running.

When I was a teen and had my laying hens in a non-rotated evil dirt chicken yard, I had a call for them to come when I took out kitchen scraps. They knew that call. The yard was about 30 feet from our back door and if I called as soon as I stepped outside, they'd come running to greet me. By the time I got to the feeding pan, they already had it surrounded expectantly.

If you really like dogs and have a hankering to train a herd dog, that can be extremely helpful with sheep and goats. In normal gait, pigs are the fastest, then sheep, then cows. Sheep are more cluster oriented than cows and goats. Even a handful of sheep move like one organism.

My grandson Andrew has a flock of about 50 sheep that he moves with a guard dog. He trained the guard dog from a puppy by putting it out in a pen by itself for several days, then bringing it to the sheep for a few days. Then he would take the puppy back to the pen. It would whimper and cry; nobody could pet the puppy. After several days, he'd put the puppy back with the sheep. In about three cycles of this, the puppy knew where he wanted to be: with the sheep.

This training regimen teaches the dog that the only place he's happy is with the sheep; otherwise, he's alone and forlorn. This dog/sheep bond is wonderful. Our farm is not fenced for sheep; we only have single strand cow-height electric fence. Sheep can go under it easily. When Andrew started with his sheep half a dozen years ago, we knew we did not want to re-fence the whole farm to accommodate sheep.

With this dog, Andrew can lead his flock of sheep from one end of the farm to the other, by himself. The sheep follow the dog. Andrew normally keeps them controlled with regular Premier sheep netting but often needs to move them long distances. Because it's a small flock, he uses them like a glorified weed eater around the outbuildings, along the fence lines, and along field edges.

He puts the dog on a long leash and calls "Here sheep." He uses the same voice and same call every time. With the dog on the leash and the call on his lips, he can lead that flock anywhere without any help. Down the road, across the road. It's pure poetry. The dog lives with the sheep at all times. Although we hear coyotes routinely at night, we have yet to have a single coyote attack. Listening to the yip yip of the coyotes and the deep responding bark of the guard dog is music to the ears of a sheep farmer.

My other grandson, Travis, has Khaki Campbell ducks for egg production. He keeps them in an electric net and they are much like sheep in their clustering instinct. Ducks do everything together. They even lay all their eggs within about 15 minutes of each other. They are actually quite easy to herd and exhibit the same flight zone pattern as cows.

He handles them similar to the layers in the Millennium Feathernet by using two contiguous electric nets. He erects one net adjacent to the one the birds are in and then opens up both of them at a throat and lets the ducks enter the new circle. After closing the new circle, he pulls up the old circle and re-installs it ahead of the new circle. This circle leapfrogging keeps the ducks contained at all times, even when they're moved.

The main thing to remember when moving, handling, and sorting animals is each one has a flight zone and it's sized a little different from individual to individual. Pursuit stresses all animals; a pulsing between pressure and rest, pressure and rest, is the way to keep things calm and heading the right direction.

Chapter 21

Alternative Feeds

Many homesteaders want to either reduce or eliminate grain feeding. Whether it's inability to find GMO-free feeds, a desire to eliminate corn and soybeans, or more on-site generated feed stuffs, this objective can lead to interesting activities. This chapter will focus on omnivores since the herbivores only eat forage. Here we're going to zero in on grain replacements or minimization.

To be sure, on a homestead with few animals, your own garden and kitchen wastes can supplement pigs and chickens. About the only grain crop you can actually grow efficiently on a micro-scale is corn. The big ears can be dried on the stalk, harvested by hand, shucked as they're harvested, and fed to chickens or pigs. All the other grains are much smaller and need to be flailed and threshed to separate the kernels from the husks. That's not an easy task.

All grains, including corn, are best fed cracked, ground, crimped or crushed. Digesting whole grain takes a lot of energy. In the case of pigs, chewing corn even takes a lot of energy. A pig's teeth aren't really capable of handling whole dry corn kernels until it's at least six months old. My goal in this chapter is not to discourage you from trying some alternative feeds, even though that may be the perception. No, my goal is to think through all the related issues in order to keep you from frittering your time and energy away on nonsense.

The more difficult accessing the nutrients in the grain is, the less digestive and productive performance per pound fed will be. The point here is that growing the grain is only one part of the equation. The other part is getting the grain broken up enough to give the animal efficient access to its nutrition. Let's assume a quarter acre of corn produces 11,200 ears. If you shuck as you pick, allowing 10 seconds per ear, you could harvest those ears in about 18 hours.

The next step is to get the kernels off the cob. You can still buy hand-operated corn shellers new, or find one at a farm auction. These are kind of fun to operate, actually. You crank the handle and drop the ears in; wheels scrape off the kernels and spit the cobs out separately. Assuming a quarter pound of kernels per ear, these 11,200 ears will give you about 2,800 pounds. If you can put 12 ears a minute through the sheller, that process will take you about 15 hours.

After 30 plus hours (15 + 18), you have 2,800 pounds of corn. Now we have to grind it or crack it. Is this starting to sound like the Little Red Hen? Whether you invest in a small mill or take it to a neighbor who has one, you'll have half a day in this process even if you take it to a neighbor with a tractor PTO-powered grinder-mixer. Remember, you need a vessel to hold all this corn. At about 7 pounds per gallon, you'll need about 400 gallons of capacity or 54 cubic feet. Imagine seven 55-gallon barrels.

After another 4 hours, you have cracked or ground corn. I'm lumping this altogether for simplicity's sake, but as a practical matter, you'd never want to grind it all at once because opening the kernel makes it start to deteriorate. You don't want to grind more than a month's worth of usage at one time. That adds a new wrinkle. If you're going to keep all this corn around, you need to protect it from mice and rats.

The most unwelcome animals on a homestead are mice and rats. That's why one of your most valuable homestead animals will be a couple of aggressive mousers, otherwise known as cats. And yes, if you set up an LLC for your homestead, all cat food and related expenses are legitimate business expenses for vermin control. Cats are just as valid as poison, that's for sure. Allan Nation used to tell me that the single most important reason for everyone to have a business is to enable us to enjoy our hobbies with before-tax dollars.

The bottom line is that in our corn example we've invested around 35 hours of time to get 2,800 pounds of corn. If our time is worth $20 an hour and we say half of this is fun and games because we really aren't in business, then our labor to do all this is worth $350 ($10 an hour X 35 hours). But wait!--you know those TV commercials. "But wait! If you order right now . . ." We forgot something. We didn't put any time into growing the corn, buying the seed, prepping the ground.

Let's fix that oversight with another 5 hours for growing the corn. I'm being extremely generous here, assuming you have a neighbor come over and prep the ground and you have a handy dandy seeder that can get it done quickly. You'll have to keep weeds at bay and you might need to go along with mineral oil to put some drops on each ear silk to keep the earworms out. This puts our total investment up to at least 40 hours. Using our $10 per hour rate (since half is fun and games) we have $400 in labor for 2,800 pounds of corn. Surely we have $200 in containers, cats, custom grinding or tilling, seeds, etc. Now we're in at $600 for our 2,800 pounds of corn.

Anyone familiar with small-scale production knows I'm being extremely generous with my numbers. I'm not trying to be prejudicial. If we divide our $600 in labor and expenses into our 2,800 pounds of corn, we're at 21 cents a pound. For the experience and sheer joy of it all, that's not a bad investment. Just realize that you can buy that corn for about 10 cents a pound.

Now you see, dear folks, why I'm an advocate of mutual interdependence. Allan Nation used to quip "growing a little grain is like being a little bit pregnant. You either are or you aren't." In other words, you're either in the grain business or you aren't. It's hard to dabble in the grain business. With mechanization and even rudimentary efficiencies enjoyed by bona fide crop farmers, a homestead grain production model is hard to justify.

To be sure, if you want to grow some grain for your own personal consumption, for baking, that's another story. One bushel of amaranth or ancient wheat goes a long way. We grew wheat in one of our garden sections one year and got nearly a bushel. It was interesting. Notice I used past tense. Been there, done that. We have great local producers and a good stone ground water-driven mill in the area that enjoys our patronage. You can diversify to the point of inefficiency. You won't do homesteading a disservice by leveraging the skills of neighbors who do things better than you. Some things are better left to others.

Remember, the exercise we just went through was for corn only; nothing else. And corn is by far the easiest to do by hand. The smaller grains are far more tedious and inefficient. If you don't have any experience in this, you can be pulled off by some altruistic sales pitch, or some purist doctrinaire, and get mired in frustrating activities. To quote another of Allan Nation's famous lines: "you're known as much for what you say no to, as by what you say yes to."

A homestead has enough interesting and exciting things to do without frittering away valuable time and money on an ephemeral paradise. Procuring your grain from folks equipped and educated to produce it is the place to start. If you find yourself vegging on the porch with nothing to do in a couple of years, then you have my full blessing to tackle some small scale grain production. Ha!

Tubers like mangels, sugar beets, turnips, and even potatoes can offer a wonderful supplement for either treats or self-harvesting. I mentioned earlier the fall many years ago when we fattened a couple of pigs on Jerusalem artichokes in a barn corral. I also mentioned our mangel trial and how the weeds overtook them.

Anything you can plant that an animal can self-harvest is a game changer. As you could see in the corn example above, the big time is in harvesting and preparing for animal acceptance. Historically, tubers provided lots of opportunity for hogs because they didn't have to be harvested or stored. The ground provided safe keeping and animal access could be controlled with fencing to apportion the crop. For the same reason we rotate pastures, in a self-harvesting situation we rotate through the crop to keep things fresh, get in and out, distribute manure more effectively, and help the animals be more efficient in their harvesting.

The definitive book on formulating rations is Morrison's *Feeds and Feeding*, which you can find in many editions. Mine is edition 22 published in 1956. It gives nutrient values of an endless array of materials. Using simple ratio math, you can create various formulas based on what you have. Table 1 in the appendix is titled "Average composition and digestible nutrients."

The nuances are unbelievably fine. For example, starting out at the top of the table with dry roughages, here are the entries, in order:

Alfalfa hay, all analyses
Alfalfa hay, very leafy
Alfalfa hay, leafy
Alfalfa hay, good
Alfalfa hay, fair
Alfalfa hay, stemmy
Alfalfa hay, before bloom
Alfalfa hay, 1/10 to 1/2 bloom

Alfalfa hay, 3/4 to full bloom
Alfalfa hay, past bloom
Alfalfa hay, barn dried, good
Alfalfa hay, brown

For each of these, it gives 17 measurements:

Total dry matter
Digestible protein
Total digestible nutrients
Nutritive ratio
Protein
Fat
Fiber
Nitrogen-free extract
Mineral Matter
Calcium
Phosphorus
Nitrogen
Potassium
Protein digestion coefficients
Fat digestion coefficients
Fiber digestion coefficients
Nitrogen-free extract digestion coefficients

Whew! I've referred to this book many times over the years when trying to figure out what a pound or bushel of something is worth. Just for fun, here's a section out of the Fs:

Feather meal
Fermentation solubles, dried
Feterita grain
Feterita head chops
Fish-liver oil meal
Fish meal, all analyses
Fish Meal, herring

Fish meal, menhaden
Fish meal, red fish
Fish meal, salmon
Fish meal, sardine
Fish meal, tuna
Fish meal, white fish
Fish, soluble, dried
Fish solubles, condensed
Flaxseed
Flaxseed screenings
Flaxseed screenings oil feed
Garbage, hotel and restaurant
Garbage, municipal

Oops, I went into the Gs a bit, but you can see the breadth of this. This table goes from page 1,000 to page 1,069. I'm putting this in here not because I'm going to formulate feeds for you, but so you'll appreciate the kind of information that's out there. These old books have a lot of wisdom and information and you can use them when the power is out; you don't have to turn them on, reboot, or subscribe to an industrial internet platform. One of the first things a successful homestead has is a library. These old dusty tomes smell as good to me as well-made compost. Yummy for soil worms and book worms.

My grandfather had an octagonal chicken house on the corner of his magnificent garden and he always kept a couple dozen hens. He devoted one of his garden beds each year to mangels and would skewer one a day on a nail on the wall of the coop. The chickens would peck at it for fresh vegetable material in the winter when fresh material was scarce. I don't know that it displaced much feed, but it sure provided the chickens with hours of entertainment and I'm sure it was a tonic for health due to its fresh food potency.

One of the reasons grain is the default ration for omnivores is because nothing else comes close in nutrition per pound. Think about how many ounces of steak is required to make you feel satiated versus ounces of lettuce. In the animal feeding realm, we have to tote this feed, either by hand or by vehicle.

While more succulent plants like cabbage and mangels are a great side dish, if you will, production per pound is way higher with the concentrated grains. They don't have much water for one thing, but they simply concentrate more nutrients per pound. A chicken could eat grass all day but not satisfy her energy needs. Grain is easy to store and it's stable. It doesn't go bad in the heat or cold. It's easy to handle mechanically with augers and conveyors.

As an experiment one time many years ago, I fed chickens a 100 percent comfrey diet. Comfrey hit the homestead literature like a race horse back in the late 1980s. We put in a nice big bed of it like thousands of others, viewing it as almost a miracle plant. What could be better than a perennial leafy soybean substitute? A year later, the same journals started carrying warning articles about carcinogenic side effects if you fed too much comfrey. In our own experiment, the chickens developed tumors.

I never repeated the experiment, but I really think if I had used the comfrey as a side dish and just corn as the main dish, they might have gotten along okay. They were on pasture too in a rotated yard. Most green and succulent feed stuffs are high in minerals and vitamins, but low in energy. In the case of comfrey, it has a lot of protein--dried it's higher than soybeans. Animals need energy to produce eggs, milk, and grow their own muscles.

An omnivore differs from an herbivore in its digestive and metabolic capacity to assimilate energy from watery material. We ran into this when a lady just four miles away put in a craft cheese facility and approached us about taking her whey. Folks, if you want to see a happy pig, feed it anything from a dairy. Pigs and dairies go together, like the farmers in the Swiss Alps have known for a long time. We brought the whey home and fed it to the pigs but even though we only ran four miles over and four miles back, we could not make it work financially.

I just talked to a friend who raised some experimental pigs last year on whey from his artisanal cheese shop. They drank 10 gallons per day per hog and still ate grain. Because the pigs were just outside his cheese room, he didn't have any transportation so the experiment worked financially and he did indeed offset a fair amount of feed costs. It worked due to the proximity of all parts of the equation. On the surface, waste feeding and gleaning seems like a no-brainer good thing, but it's not as easy as those of us who like to divert all edible wastes away from the landfill would like to think.

Gleaning is an umbrella term for using food wastes and of course for salvaging field edges, blemishes, and anything else edible by animals that humans won't eat. If you're like me, you hate to see those dumpsters full of discarded produce coming out the back end of supermarkets. Right now, worldwide, about 40 percent of human edible food is never eaten by a human; most of that simply goes into compost or landfills.

My heart is completely on board with diverting food wastes to pigs and chickens. But this has two problems. The first is that much of our food is junk and not healthy for animals. If you wouldn't eat Little Debbie's Ho Ho cakes because they lack nutrition, why would you expect your animals to eat them and stay healthy? We tried gleaning from the local food bank years ago and the pig meat went soft and bland. Our heart was in the right spot, but it was well-intentioned folly. Now the food bank sends mountains of stuff to the landfill. Of course, a lot of that donated food isn't fit for human consumption anyway.

As the co-owner of a slaughterhouse (T&E Meats in Harrisonburg, Virginia) I'm privy to feedback from the butcher crew. Handling carcasses from more than 100 area farms, these craft cutters see the nuances from one person's stock to another.

Invariably, farmers who feed gleanings to their hogs have the poorest meat quality. It's flabby and not firm; it lacks tone and body. The meat cutters always complain about having to cut that pork because it's more like cutting jello than cutting refrigerated cookie dough.

The second gleaning issue is time and money. Let's assume you work with a couple of local supermarkets to take their waste produce. Each generates two barrels a day. In the winter, it freezes and gets caked with ice and snow on the dock outside. In the summer, if you don't get it every day it becomes a fly-infested smelly mess. That means a trip to town every two days, at least.

Unless you're on the edge of the city limits, you can't make a round trip to town in much less than an hour. Let's say you can make two stops and do the round trip in an hour and a half. Assume it's 25 miles total, at 50 cents a mile, is $12.50 for fuel and depreciation on equipment, plus $30 in labor ($20 an hour). We're going to cost this out at a full value because this isn't exercise and it isn't fun. You went to your homestead to enjoy the homestead, right? You didn't go to your homestead to run to town every other day.

This trip, then, costs $42.50 for the four barrels of salvaged produce. If you look at true nutritional value, that produce might be worth about 20 pounds of grain feed per barrel. You're hauling a lot of water and not much energy. Four barrels at a 20 pound per barrel equivalent is 80 pounds of feed value. If anything I'm being generous with this example. In relative feed cost, then, those barrels of gleanings cost 53 cents a pound ($42.50 divided by 80 pounds of relative feed value). You can get a perfectly blended Total Mixed Ration (TMR) of GMO-free local grain for less than 30 cents a pound.

But wait! We're not done. You have to feed the gleanings when you get home. It's not nearly as easy to feed as grain because it doesn't fit neatly into a no-spill feeder. You can't scoop it out neatly with a feed scoop. If you're feeding inside, how do you keep the animals from soiling it? If you're feeding outside, how do you keep them from grinding it into the dirt?

And the worst thing--been there, done that--is the inedible stuff. Paper, plastic, utensils, ink pens. You have to pick through it and chase down wind-blown pieces across your field. Your own trash barrels fill rapidly, which necessitates a quicker trip to the dump. Again, you started your homestead to be a beautiful respite from chaos, not a dumping ground for urban waste.

Normally, your animals will balance out their own nutritional needs with gleaning supplementation as long as you have a TMR accessible. A pig won't over eat on a particular item to its health detriment. Fortunately, animals have enough instinctual sense to eat carbohydrates when they need them and protein when that's what they need.

When I built my first bona fide Eggmobile and put 100 chickens in it, they literally lived off the land. I gave them some corn to make sure they had energy and they got along okay. They knew how many worms they needed to eat compared to how many kernels of corn. I'm a firm believer now that whatever gleaning you want to do, you should still have a TMR option to enable them to balance their own needs. One year in the woods we had a tremendous acorn drop (yes, *Feeds and Feeding* even gives a nutrient analysis of acorns) and we ran pigs through an area by simply zig-zagging electric fence from tree to tree, using cheap nylon rope as poor-boy insulators.

We tied the rope around the tree to locate it, then left the two ends long enough to tie a knotted loop that held the electric fence wire. The rope doesn't hurt the tree (or a future chainsaw operator) and we can efficiently encircle a large area cheaply and quickly. This particular group of about 50 hogs was in the final fattening stage and eating about 400 pounds of feed a day. We moved them into that oak glen, along with their TMR self-feeder, and they didn't touch that feed for five days. By their choice, not mine.

They couldn't have been happier. For five days I knew we were saving 400 pounds of feed a day. This is gleaning done efficiently and profitably. No hauling, move the animals to the salvage, let them self-harvest, and control them cheaply. That is the recipe for successful gleaning.

Our former apprentice Jordan Green has a large pastured farrowing operation and lives near an industrial apple processing facility. It makes apple juice and vinegar. By purchasing dump truck loads of discarded apples, he makes the hauling work because of scale. He's not toting in a couple of barrels a day; he's bringing in dump truck loads at a time, in the fall when the temperature is cooler, with a specific commodity that fits his sow ration. Since he's limit feeding the sows and has to visit them every day anyway, carrying apples in addition to the grain does not add anything to the feeding trip. Even this is still a delicate dance to keep pork quality acceptable.

I can feel the push back through these pages: "what about the pesticides on those apples?" Yes, that's one of the considerations in gleaning. Remember, chlorophyl is a detoxifier, so as long as the pigs have access to lots of green grass, that will bind and neutralize a lot. But would you rather landfill these apples, completely removing them from the carbon cycle, like the plant was doing before Jordan came along?

Peanut factories often have sweepings and discards. Again, we wrestle with the chemical issue. Ditto scrounging up all the left-over pumpkins from Halloween. Hogs love pumpkins. Might they have some chemicals on them? Yes. This is another reason I don't participate in the government organic program. Letting perfection keep you from doing good, in my view, is not the best way to move the needle incrementally to where we want it to go. I'd love it if we eliminated putting edible feed stocks in landfills; or compost piles for that matter, although that's better than landfills.

Perhaps one of the nation's most successful gleaners is Vermont Composting, which uses food wastes to feed a couple thousand chickens to stir a massive compost pile to grow worms to feed the chickens to make one of the best potting soil mixes on the planet. Much if not most of those food scraps are from industrial sources, from sources some of us would not patronize. But by tapping into the synergistic protections and alchemy of worms, bacteria, chickens, and oxygen, magic happens: one of the best growing mediums you'll ever find. And the eggs are to die for. Incrementalism is okay.

I had a lady call me from North Carolina who lives adjacent to a major yoghurt manufacturing facility. She said they dump a tractor trailer load of milk every single week. The reasons vary from sell-by expiration to antibiotic residue. I told her if I lived where she did, I'd be in the hog business tomorrow raising milk-fed pork. These opportunities exist, but in order to work all the pieces of the puzzle have to fit. And I wouldn't spurn the milk because it's industrially produced. Take what you have and complement it to something better.

Will Allen, Milwaukee's waste stream worm/fish/chicken icon, turns mountains of food waste into nutritious off-the-chart food by running it through an alimentary canal (that's worms, for the uninformed). The guy is a genius; when I grow up, I want to be as smart as he is. Does he turn down food scraps grown with chemicals? No. He depends on the worms to make all the adjustments.

For years we've tried to figure out a way to bring back restaurant discards from the 50 restaurants we service. Why can't we go loaded and come loaded? But for some reason, chefs don't like to pull their chicken and eggs out of the truck past stinky kitchen scraps from the competitor down the street. We haven't cracked that nut. And we haven't figured out how to send another vehicle on a gleaning run to all these establishments and make it pay. Gleaning can work, but you have to juggle a lot of variables, which I hope I've addressed in this little discussion.

Another alternative feed stock is fodder systems, or sprouted grains. A tremendous amount of research and technology is being devoted to creating turn-key outfits. The first trials I did were here in the back yard. I put a couple of 2X4s on two barrels and used seed germination trays as containers, sitting on the 2X4 stringers. I tried different grains and different mediums: compost, newspaper, soil.

Almost all of these systems use a 6-day sprout cycle. The biggest danger is mold and fungus. I did mine outside in the summer in a poor-boy set-up that kept things rustic enough to keep nasties from invading. The more sophisticated and controlled your environment, the less overall ecological balance the system has and the more susceptible to nasties.

My experiment went well and the chickens loved the biscuits (that's what the mat of sprouts is called) but it didn't seem to greatly change what they ate at the feeder. True, maybe if we had kept it up and done a more comprehensive and long-term study we would have seen some taste or fat profile differences. I'm a fan of sprouts, philosophically if not practically. A few years later we had a sprout company bring in their computer-controlled state-of-the art sprout trailer but again, it didn't seem to make a big difference in feed consumption.

Unfortunately, both of these trials occurred in the summer, when the chickens also had grass. We have not done any sprouting in the winter. My hunch is that's where big savings and big differences might appear. But of course that's harder because the grains won't sprout unless they're warm. That's what these completely enclosed, environmentally controlled outfits offer. I continue to be interested in sprouts and maybe in a few years, if I were writing this section, I'd change my tune to "folks, you've got to do sprouts."

As they say in my Virginia back woods, "I ain't fer it nor agin it." I do believe that at a small scale, some bootstrap sprouting can be a wonderful tonic for your pigs and chickens. Probably winter is far more important than summer, when you already have green material. But nothing I see in sprouting makes me think it can replace a TMR main dish. I hope that's fair and keeps the door open to more sprouting experiments.

As Harvey Ussery demonstrates, at small scale you can do some amazing supplementation with worms, soldier fly larvae and other fringe things. One of our former apprentices drilled a bunch of holes in a 5-gallon bucket and hung it out with the chickens as a mortality depository. Road kill, dead chickens, occasional bureaucrats--you know, anything with meat--he'd throw in that bucket and attract flies. They laid eggs; the eggs hatched into maggots, and the maggots fell out of the holes onto the ground, whereupon the chickens feasted and made exceptionally good eggs.

He only discontinued it when he couldn't stand the smell. I saw an article about a guy who used the same idea, but he floated the bucket on his pond to feed the fish. I'd like to do that; if I do, I think I'll call it a "Grub Tub." Oh, so many things to do and so little time. Story of my life.

We tried an earthworm bed under the rabbits. The bed fit exactly under the rabbits and we made 4-foot wire cloches to exclude the chickens. Each day we removed one segment and let the chickens scratch into the worm bed and self-harvest worms. We even incorporated a rye sprouting part of the cycle. It worked great until really cold winter set in; then the worms moved out and the rats moved in. What a mess. We could have filled lots of Grub Tubs with dead rats in the spring when we tore apart the defunct worm beds. You want to generate excitement on a homestead? Get a village of rats, then tear it apart and chase them down with scoop shovels, pistols loaded with rat shot, and silage forks. Talk about red-neck entertainment. Your city cousins would pay big bucks to be part of that action.

Are you tired of our failed experiments yet? Don't be discouraged. All of these have points of opportunity. The trick is to dust yourself off from failed experiments, do them small enough that they don't sink the mother ship, and live to try another day. That's the never-ending hope and eagerness of an innovative homestead. May you ever be in experimentation mode.

Chapter 22

Goats and Sheep

Full disclosure: right now we don't have goats at Polyface. One of our former apprentices who has come back to join our team full time has a small flock of goats on one of our rental farms. I haven't personally raised goats for a long time. But we did. And it looks like they're re-entering my life this moment.

We had goats long ago. I remember the day we decided to divorce them (you really do marry goats, you know). When I get up in the morning, I walk over to the window to see what the weather's like. We park our cars out front, in a direct line of sight from our bedroom window. I looked out and saw four goats in the ecstasy of cud chewing lounging on the hood of the car.

I got dressed, went out the back door, and encountered another goat halfway up a ladder I'd left up against the house the day before. I hadn't quite finished cleaning out the gutters and left it up overnight to finish the next day. The goat was 10 feet up the ladder and looked at me like "thanks for the playground, human. What will you do for me tomorrow? This is pretty cool."

I went on out and did chores, then came back in and announced that the goats were leaving. We signed the divorce papers before the week was out and I have no idea what kind of life they lived after they left. I cared, but not enough to continue correspondence.

Okay, all you goat lovers, let's get serious. Enough jokes at your expense, and thank you for indulging my incompetencies. Here is the most important husbandry truth about goats (besides that you need a good fence): they are browsers, not grazers. Sheep and cows are grazers, meaning they tend to eat **below** their shoulders. Certainly you'll see both cows and sheep reach up and eat things above their shoulders from time to time, but for the most part, they want their heads down.

Goats, on the other hand, are browsers, meaning that they want about 80 percent of their diet coming from **above** their shoulders. This is such a critical difference that everything about their care revolves around it. Their entire habitat depends on this preference.

Even goatherd novices know that goats love to climb bushes, embracing woody vegetation with their legs while nipping off leaves with lips that literally flutter with activity. I know they don't move as fast as a hummingbird's wings, but if you could put a lip speedometer on their mouths, they'd peg the needle. You can see goats grazing, but not nearly as often as browsing, if they have browse available. Therein lies the great conundrum of goats.

Goats prefer woody plants to forages, about 4:1. But woody plants require twice the rest time compared to forages to recuperate from defoliation. In other words, if a grazed grass plant needs a month to grow back to grazing height and root energy replacement, a woody plant in the same environment would need two months. I'm being generous; some would say it should be three months. For now, let's not get bogged down on the time; let's just agree that woody plants require far more recuperation time, all things being equal, than forages.

This is why the gift of goats to the homestead is forage improvement. Historically, in nomadic cultures, goats prepared the vegetation for sheep; sheep prepared the forage for cows; cows deteriorated the forage for the goats. In the plant succession, goats kill the woody species and weeds, encouraging the grass to grow. Sheep, which have a broader palatability index than cows, continue this forage succession to yet another level. Cows, the queen of livestock and possessing the narrowest palatability index, select the highest quality forages, gradually diminishing them.

As the cow-delectable species gradually give way, nature covers the ground with plants cows don't like as much. First come weeds and then come woody plants. Then it's time to bring in the goats and repeat the process. That's a historically normal cycle from around the world that is still evident in many cultures. To be sure, proper grazing management like I've outlined in this book reduces this pasture deterioration. Exceptional pastures are almost impossible to maintain without a complementary species or the mowing machine.

A permutation on that diversity theme, of course, is mixed species grazing, where all three species stay together in a *flerd* (flock + herd = flerd). In that scenario, the ratios are critical. I don't know what they are, and I suspect they would be different in different regions. I also suspect that barring indigenous wisdom on the subject, it would take about a lifetime to re-discover what those functional ratios are.

When I toured the Delahesa region of Spain, famous for its black-footed acorn-finished hogs, I was fascinated by the intricate ratios of cows, sheep, and pigs to maintain the pastures. On one farm, the owners told me I was standing under a 500-year old tree and the landscape had not changed significantly in 1,000 years. As any student of ecological systems knows, a functional eco-system that can perpetuate with little change for 1,000 years is extremely stable. Indeed, it's the ultimate holy grail of agriculture, is it not, to create a highly productive, self-sustaining system that continues luxuriantly 1,000 years without deteriorating.

Unfortunately, in recent years as the coveted pigs and their 5-year seasoned meats gained market recognition, these historical sacred ratios no longer hold. Farmers eager to cash in on their pork popularity have abandoned sheep and cattle in favor of hogs, many of which are now grain-fed, and run in numbers that destroy the pastoral park-like landscapes punctuated by magnificent hand-pruned and trained trees that produce the acorns. I was chagrined to see torn up pastures and trees struggling against overly aggressive disturbance and not enough rest periods between pig visits.

Diversity is not some pie-in-the-sky academic subject; it is real with practical implications. I would suggest that the biggest problem in goat rearing is trying to turn these marvelous, resourceful animals into grazers. If you're following the train of thought, realize that goats de-vegetate the brushy plants first. Goats leave clover to tackle the bushes. This opens up the canopy for the ground surface vegetation like grasses and clovers, to get a toehold. Growing unmolested, these forages gain ground, literally covering more ground and becoming more robust. None of this is exclusive: goats will certainly eat some clover and some grass before they completely defoliate the weeds and bushes, but they certainly do not eat all the clover before eating the bushes. It's the other way around. The forages recuperate faster both because they weren't grazed as much and because they now have more light through the canopy.

Seeing all this forage, the goat farmer turns the goats back into the area, but what the goat really prefers are the woody plants, which have not fully recovered from their previous pruning. The goats attack these woody plants that require a longer recovery period, denuding them again. Suddenly, these woody plants and weeds find themselves at a distinct disadvantage to the ground covering forages, which of course exploit this new ecological niche opportunity and increase strength.

As this cycle continues, the woody species actually die and the forages ascend to dominance in the vegetation canopy. This is a problem for the goat owner because the goat prefers to eat above its shoulders. When the goats have nothing to eat except what's below their shoulders, it sets a whole series of health problems in motion for the goats.

The nutritive value, the punch if you will, of the forages, and especially the protein, is higher than the woody plants and weeds. The woody plants are lower in energy, lower in protein, but higher in minerals because their roots go deeper. Many of these weeds and shrubs actually grow in a place specifically to find and bring to the surface deficient minerals. Forages tend to capture minerals that are available on location, not rectify deficiencies. Weeds and bushes, on the other hand, seek, transfer, and rectify these deficiencies.

The goats, then, who had to work harder eating above their shoulders, bending over fronds and branches, and doing what goats do, suddenly find themselves eating candy bars and ice cream. It's too rich a diet for a goat. That sets a degenerating health trajectory in the goat: overgrown hooves, retained placentas and other birth problems for oversized kids, digestive disorders (diarrhea) and other things. These common goat maladies are almost always the result of turning goats into grazers when they prefer to be browsers.

That's why we often say "goats eat themselves out of their habitat." In other words, by improving their landscape to forages, goats create a less hospitable habitat in which to thrive. This is the goat conundrum because often it takes about five years for this cycle to run its course. Generally, it takes about five years for newbies to begin loving their goats.

In other words, at the five year level the new goat farmer finally figures out how to keep them in and off the car hood. She (lots of goat farmers are women) figures out the breeding cycle, learns to abide the noxious buck and all his idiosyncrasies, and develops a market for chevon or goat milk. This all takes time and the possibilities of mastery finally start to shine at year five. But that's exactly when the goat destroys its favored habitat.

What to do? Indigenous nomadic cultures solve this problem by maintaining either a rotation (goat, sheep, cow) or running a tight ratio flerd that offers all the species a desirable option of plants. This is where I say what some folks don't want to hear: a single species outfit rarely works. That includes goats.

The reason this problem is most acute with goats is because you can't rectify the deficiency with a mower. If cows gradually deplete their succulent forages, allowing weeds and woody species to encroach, we can fix it with a mower. In many cases, a mower is cheaper than a goat. But you can't use a machine to rectify the goats' damage to the woody species. You can't plug branches and leaves back onto those plants.

The most successful long-term goat folks I've seen are people who run a flerd and settle on a sustainable ratio, like one goat per ten sheep. That way the goats get the weeds and brush while the sheep eat the low stuff. Unless you have an extremely weedy field, you'll never want to run them 50/50. But you always have a few weeds or brush encroachment, so the goats at a low stocking percentage handle that while the sheep or cows handle the forages.

Will goats eat forages? Of course. Don't you like ice cream and candy bars? In this case, if the animals could talk, they would tell us "we want our brush back." But animals always aim to please, so they go about their business and smack their lips 1,000 times per minute. We might think they're saying "thank you very much" but in reality, they're saying "we wish we had our browse back."

The flerd idea is easier to do the closer aligned the species you put together. Sheep and goats are extremely compatible. Adding cows to the mix means you need more weight on your water tank to keep the cows from knocking it over and you need more height on a shade structure to keep the cows from rubbing the top of it. Because these are all herbivores, putting them together is relatively easy.

The flerd idea is harder when you start mixing herbivores and omnivores due to grain requirements for the omnivores. Somehow you have to exclude the herbivores. That can certainly be done with a small door that only a chicken can enter, but you'll also need to exclude the goats from the nest boxes. A totally enclosed chicken feed and nest box area, like a compartment within the larger netting corral, will work as long as your access door is small enough only for a chicken.

If you have lambs or kids, though, you can't get the door small enough to exclude those babies. See the problem? You can only chase the flerd idea so far before it actually becomes too cumbersome. As you push incompatibility, the flerd idea starts to break down.

If you add pigs, you'll need to exclude goats and sheep and cows from the pig feed. Fortunately, any watering set-up that will service a pig can service any of the other animals. But keeping a lamb out of the pig feeding area can be problematic. The singular advantage for a pig is its ability to lift things with its snout. A guillotine door with some weight can work to let in pigs; it has to be fast acting, though, or the clever goats will figure out how to squeeze in with the pigs and eat the pig feed.

In general, I think the incompatibility of herbivores and omnivores is a big enough issue to run them separately. Obviously an adjacent leader-follower set-up can work fine. That offers the benefit of multiple animals in one netting set-up but protects you from the frustration of trying to exclude omnivore feed from the herbivores.

The other extremely successful management response to this problem is to start a goat clean-up operation in your area. Rent-a-Goat and other schemes are highly effective at vegetation control off your property. This can generate another income stream, whereby people pay you to run your goats on their property. Many small goat operations offer this service and it's an excellent way to keep the goats on their preferred vegetation, doing the work they enjoy most.

As for fencing, goats are certainly the most difficult of all the farm animals to control. In my experience, they are not too difficult if they aren't pushed. Remember, your goal with goats is not to prune off the forage like the cows do. Your goal is to move through an area and let them eat the brushy plants and weeds, then move on. In such a case, they can always eat some grass if they get hungry. The amount of feed on offer keeps them from ever getting hungry.

As long as they are content, they don't push the fence too much. But if you start pushing them hard, demanding that they eat not only the brush but all the grass too, their discontent will spill over into a desire for adventure and that inevitably leads to frantic screams "THE GOATS ARE OUT! THE GOATS ARE OUT!" And when they head to the garden's tall tomato plants, or the green beans, a couple of wayward goats can do a lot of damage in a few minutes. Keep them content; you don't have to push them hard because they'll naturally eat the bushes and weeds first.

If you don't like netting, three stands of electric fence will also do the job, again, as long as they aren't pushed hard. With all that said, one of the beautiful things about goats is that since they do eat primarily above their shoulders, they are the one herbivore most conducive to long paddock stays. Cows and sheep graze across their dung. But goats don't. That means from a sanitation standpoint, you can leave goats in an area for several days and not be as religious about moving them frequently.

The silver lining on the fencing, then, is that while they do require the most perfect fencing set-up, you don't have to change it as frequently, either for their health or for the ecology. The bushes and weeds don't grow back very fast anyway. Unlike forages, which in a fast growth time can grow an inch a day and therefore suffer the insult of a second bite, or second pruning within a day or two, weeds and brush in their retarded recovery don't pose that problem. A one week stay in a given paddock for goats is perfectly fine.

Goats don't need much water; they love dry weather. You can also put goats on a run, like a dog run. The problem is that if you're running them where a bunch of bushes and weeds are, it's almost impossible to keep them untangled if they're on a leash. Tethering to a stake is virtually impossible. But an overhead run, stretched between tree branches, for example, can sometimes work if the vegetation isn't too dense.

Goats are wonderful animals managed in their niche at the right time for the right objectives. They can be incredibly beneficial but frustrating at the same time. Remember, you'll start loving them at about five years. Don't give up too soon.

Now to sheep. The large wool breeds are most like a cow and the smaller hair breeds are more like a goat. With the explosion of sheep breeds in recent years, shepherds enjoy a wide array of sheep phenotypes and function.

In general, cows like a climate warm and wet. Sheep like a climate cooler and dry. Sheep thrive in high deserts; they're just a more naturally arid and rugged animal than a cow except for fighting off predators.

In most parts of the U.S. now, nobody would seriously contemplate sheep production without also having a guard animal. The most common is a dog but it could also be a donkey or alpaca. For most, a dog is the most familiar option; it requires the least amount of new learning. Dogs are far more common in our American society than donkeys and alpacas. Finding dog mentors is easier than finding donkey or alpaca mentors.

On a homestead, on a small acreage, any kind of good, disciplined, watchful dog will work. I'm a fan of farm dogs that patrol and kill things. We have perpetual war on groundhogs and any dog that seeks and kills those things is a friend of mine. One time I had the privilege of watching one of our long-time guard dogs, Michael, catch and kill a groundhog. It had to be one of the most violent things I've ever seen, but it was also exhilarating to see this dog rid our field of a cow-leg-breaking and wagon-axle-breaking critter. Not all critters are born equal.

One of our former apprentices is doing an experiment right now trying to get sheep to bond to cows. It's a small flock of sheep (about 15) and the cow herd is 200. The sheep naturally stay near the cows and at night, go in among them to sleep. The cows offer natural protection for the sheep. Although the sheep can go out under the cow wire, they don't stray far. This cow-sheep bond is definitely a relationship worth developing, but I don't know if in order to be successful it requires a certain ratio. We tried to duplicate it with Andrew's much larger flock and it was unsuccessful. These kinds of relation-building schemes demand the highest skills of anything you'll ever attempt.

In general, any electric fence that will hold sheep will also hold a cow. The cow-sheep option I think is one of the most efficient options for a homestead. If you only have a couple of cows and a handful of sheep, you can enclose all of them in either a net or a three-strand fence and almost forget any permanent fencing whatsoever. All you have is a laneway for access and your open fields. In this kind of set-up, you move the flerd without regard to permanent fences. I call this the ultimate self-contained system. This works just as well with goats, or you could add goats to the sheep and cows.

With small groups, the more you can group, the less fence you have to put up and take down. Efficiency increases when instead of running the cows in a separate paddock with a single wire and moving the sheep in their independent net, for example, you combine them in a net and forget the single stranded cow fence.

With my grandson Andrew's flock, we've created some efficiencies by doing a leader-follower program with the Millennium Feathernet. This is a hybrid idea of the multi-species clumping option. Rather than putting them altogether, we run them separately but use one netting set-up for both animals. Chickens like short grass. We set up a quarter-acre oval for the sheep; they graze there for three days. Then we move the sheep to their next oval and bring the chickens to the vacated sheep oval.

At that point, the grass is nice and short, which the chickens love and predators hate, and we have no additional fencing set-up for the chickens. Each net set-up gets six days of use: three days with sheep and three days with chickens. This model does require an extra set of nets, but that's a small price to pay for doubling up the use of each set-up. We've thought long and hard about how to combine the sheep and the chickens in the same net, but haven't figured out an efficient and clean way to do it.

The sheep would need to be excluded from the nest boxes and chicken feed. The current roost set-up in the scissor-trusses of the A-frame shelter would need to be changed to accommodate the sheep entering. It's complicated. So far, the best we've come up with is a leader-follower arrangement that leverages the netting set-up and amalgamates chores to the same spot on the farm. That leverages the going-and-coming getting-there trip.

If you're into fiber production, you'll certainly want wool sheep. But if you don't have a value-added market for the wool, shearing will probably be a net loss. The price of wool and price of shearing sometimes equal out. You can certainly learn how to shear and invest in the equipment to do it. You may have to discard the wool as a waste product unless you can find a buyer. That process is much harder today than it was 50 years ago. Many shepherds now use hair breeds due to the low value of commodity wool.

We've addressed parasites in sheep already; they're the number one problem in the industry. We believe, and Greg Judy concurs, that this is as much a genetic problem as a management problem. Moving them frequently and allowing plenty of rest between exposures to the same ground are key to breaking parasite cycles. Beyond that, though, is a strong genetic propensity toward susceptibility or resistance. Commercial growers are often more ruthless in their culling than a homesteader with just a few sheep.

I realize that letting two sheep die from a flock of six is disheartening. Somebody like Judy might respond, "seeing 50 die of 100 is also disheartening." But he did it and lived to tell the story. Now he has one of the most parasite resistant flocks in the nation. The easiest advice I could give you is the hard-nosed "just let the weak ones die. They'll sort themselves out and pretty soon you won't have weak ones." But when you've named them all and the kids each have their favorites, that's a hard sell.

Here's my compromise suggestion: never worm the whole flock. If one gets worms, treat it. One of the small flock advantages is that you can individualize your care far easier than a large-herd commercial outfit. Enjoy that level of customization per animal. Resolve to never do a blanket treatment of anything. Intervene only therapeutically and mark those treated animals for removal as soon as it's feasible. This may slow down your genetic hardiness trajectory, but it's probably more palatable on a homestead situation.

All that said, numerous natural remedies are available, and not just for worming. Herbal concoctions, feeding green pine needles. In this book I resolved not to repeat all the stuff contained in other books. Natural remedy recipes exist for almost anything. Search for them and try them. I would still suggest that you do individual treatment to maintain the overall objective of building resistance through genetic selection. The main rule here is not to use a treated animal as breeding stock. As soon as practical, cull animals requiring treatment to guarantee you hardier stock each year.

One thing we've found on our farm through Andrew's experimentation is that the wide-holed sheep and goat electric netting is far superior to the small-holed poultry netting if you're trying to contain sheep and goats. It's much cheaper because it doesn't have as much material. It's lighter for the same reason. But most importantly, it carries a better shock because it doesn't have as many wire filaments. A well-trained flock of sheep can be held in long narrow ovals and across difficult terrain with this high tech netting.

Now a word about winter feeding. Because goats eat above their shoulders and sheep below, hay feeding set-ups should honor these desires. A hay feeder that holds hay high is good for goats. They love to reach up high and pull hay out through slats. A simple X frame, with the bottom being the feet and the top being the V that holds the hay, works well. I always like to put a chaff box below the hay to catch seeds and leaves. The goats won't lick them all up, but at least all those leafy goodies won't fall directly to the floor. A chaff box at least offers a circuitous route toward wastage and not a straight fall.

For sheep, a box that they stick their heads in works well. It can be a simple box with head-sized access slats along the side to keep the sheep from too easily wasting hay. Making access a little difficult helps keep the sheep from grabbing a mouthful, backing up, shaking about, and dribbling half of it on the floor.

If you mix sheep and goats in a hay feeding situation, invariably the goats will jump into the sheep hay. Hay is too precious to waste. Make sure the goat hay is high enough that they can't jump into it. If you do feed the two together, you can make a double-decker station. The goat box on top, chaff plate, and then the sheep feeder on the bottom. The problem with this arrangement is how to put the sheep hay into the lower feeder. Perhaps a hinged panel to let the whole side swing away and then lift it to snap shut would do the trick.

I haven't built anything like this because we've never tried to feed sheep and goats together. But knowing how wasteful all animals tend to be with their feed, I'm always cogitating on how to minimize waste and maximize consumption. The point is that sometimes you can diversify to inefficiency. If you are feeding both sheep and goats hay, the easiest solution is probably to segregate them and have goat-friendly feeders for the goats and sheep-friendly feeders for the sheep.

Neither sheep nor goats ever needs grain. Grain is both unnatural and too rich for both of them. I never liked lamb until I went to Australia. The difference is that U.S. lamb is grain fed and Australian is not. Lamb is naturally more lean and clean than beef; it responds quicker and more negatively to grain fattening than beef. Just like with beef, grain fattening may speed up the process, but it certainly changes the taste, texture, and nutrient profile of the meat. Grain-fed lamb accentuates all the negative qualities of grain-finished herbivores even faster and more dramatically than bovines.

If you're milking goats, they don't need grain either. That's not to say that you wouldn't give them a handful of grain to get them to hop up on their milking platform without a tussle. But chances are you can achieve the same incentive with a bit of super hay like clover, or even a fresh-picked handful of honeysuckle or kudzu. Treats don't need to be sweet; they only need to be different, as long as the different is tasty.

One of the big advantages of sheep and goats versus cows is their size. They're more child compatible and all the infrastructure, from corrals to head gates, is miniature compared to cow size. Children are not nearly as intimidated--and rightly so--by these smaller herbivores. With the homestead movement in full swing, finding rams and bucks and starter groups is much easier than it was in the 1970s, for example. Getting in is simpler now, and that's a good thing.

Finally, let's talk about consumption. I bring this up because the annual per capita consumption of lamb is .7 pound in the U.S. Goat is far less. Beef is a bit more than 50 pounds. Again, in no way am I disparaging sheep and goats as a great homestead animal, but the problem I see routinely is a homestead that isn't matching its production to its consumption.

The whole idea of homesteading rather than commercial farming is to create self-reliance and foundational life integrity: know where your energy comes from, your recreation, your exercise, your food. The objective is highly personal and customized to life style and life need. If you don't match your homestead production to the way you live and eat, it'll lose its luster real fast. What's the point?

A mismatch creates two options: fatigue and frustration or selling the production. Certainly nothing is wrong with selling lamb or chevon. But did you come to the homestead to sell things, or did you come to the homestead to eat well and live the way you think is more authentic? Not very many homesteaders enter this dream from a marketing, commercial, sales mentality. And when we produce more than we can consume, or produce something we don't want to consume, it grates against our overall dream.

Before jumping into sheep or goats, a fair question is "does our family eat lamb and chevon?" Or if you're on the dairy end of it, "does our family love goat milk and cheese?" Obviously you can make soap out of goat milk as well. I'm not trying to dampen any enthusiasm, but I am trying to help you work through your steps of development.

Chances are, if you're a fairly normal American family, you eat chicken, eggs, bacon, and hamburgers (beef). Rare is the family who would say "we eat lamb, chevon, and goat cheese more than anything else." If that's your first answer, then I'd say goats and sheep are definitely a good fit for you. If you eat extremely small amounts of lamb or chevon like most Americans, I'd encourage you to think twice about getting them for your homestead.

Plenty of commercial farmers raise things they don't personally eat fondly. But on your homestead, your priority is to feed yourself first. Although cows may seem a bit more intimidating, if beef is your preference over lamb, then I think a bovine is a better fit for your homestead than a sheep. The difference in objectives between a farm planning to sell everything and a homestead planning to eat everything is quite significant. Don't oversell your emotions to something that's cute and miss the opportunity to really fill your plate with homegrown fare.

The second you grow more than you can eat, you enter a new world of marketing and commerce. That's not a bad world, obviously, but it's usually not the first druthers of a homesteader. You don't need to be conflicted about your priorities. Keep it simple and you'll enjoy your homestead far more.

Chapter 23

Milk Cow

Few animals are as coveted on a homestead as a milk cow. Few animals satisfy as many human food needs as a milk cow. Few animals demand as much attention as a milk cow.

A family milk cow was one of my dad's first acquisitions once we moved in and got settled. Her name was Suzie and she was a Guernsy-Jersey cross. I learned to milk with her udder. She finally got too old and we replaced her with Sally, a part Holstein. Wow, did she give a pile of milk. Then we had Crooked Tail and G2 and Gertrude and Whitey.

We made cottage cheese, butter, ice cream and of course drank milk like water. Raw milk, to be sure. We don't milk now but have a raw milk herd share from a grass-based dairy nearby. The decision to let the cows go coincided with my return to the farm. Instead of earning a living off the farm, we suddenly had to earn a living from the farm.

Since milk was illegal to sell and broilers enjoyed the PL90-492 producer-grower exemption, we jettisoned the milk cows and dove into pastured broilers. When we transitioned from kind of a glorified homestead supported by off-farm dollars to a full-on farming venture with no outside support, we had a new set of economic constraints around our activities.

In the time it took me to milk a cow, I could produce $60 worth of chickens that were legal to sell and buy the $2 worth of milk I needed every day. Had milk been legal to sell, I'm confident we would have gone toward dairy rather than poultry. My first farm business plans, as I contemplated how the farm could pay me a salary, was to hand milk 10 cows (plenty of people have done that) and sell the milk to folks in the community *at conventional retail prices*.

I've never quite gotten over the fact that food regulations prohibited me from starting where I had more expertise and literally kept me off the farm for a couple of years. That experience still fuels my indignation at being told what I can sell and what I can buy. Government officials picking and choosing my fuel for my own body is bad policy. My health, my body, my choice.

The only time that doesn't work is if society owns me, which of course is where America is headed. It's not a good trajectory, to assume society is responsible for my health. Such perceptions inevitably lead to criminalizing unorthodox foods and unorthodox health procedures. Enough of that. Suffice it to say I grew up milking cows, love family milk cows, and would have moved in that direction had bureaucrats not prohibited me by violent force.

For the record, I have not put in a chapter about the family milk goat because much of what I'm going to say about the cow applies to the goat. In fairness, the goat is cheaper to acquire, doesn't eat as much, and gives milk often easier to digest. Here are the goat negatives:

1. Similar effort but much less production. Most of the milking investment is getting to the animal--or getting the animal to you--cleaning, straining, bottling, and washing the accessories. The actual milking time (squirting) is similar but one gives a quart and the other gives a couple of gallons.

2. Keeping a goat in. Back to fencing issues. One strand of wire and a cow is happy and content and will never jump over or go under. She placidly stays in.
3. Babies don't make beef. This harkens back to the discussion at the end of the last chapter about what your family eats. A calf grows up into an animal that produces beef; a goat grows up into an animal that produces chevon.
4. Baby calves stay in easier than baby goats. If you think a full grown goat is hard to control, try a baby goat. Baby calves, however, train readily to a single strand of electric fence and stay in easily.

With that out of the way, let's wade in first to the main difference between a beef cow and a dairy cow (or a meat goat and a dairy goat): the dairy animal puts fat in the bucket instead of on its back. That may sound trite, but it's incredibly important because it means a meat animal can use body fat to get through lean times and then make what is called compensatory gain in good times.

A dairy animal is day-to-day. If a dairy animal misses a day of nutritional satiation and doesn't have enough left over energy to put milk in the bucket, she doesn't give double the norm the next day if she gets plenty to eat. A beef cow can go through long periods of weight loss and then gain it all back rapidly when forage is available. The beef animal can lose weight occasionally and gain back extra later.

Dairy animals need their full nutritional needs met every day in order to put something in the bucket. A milk cow doesn't let you make big mistakes. She's a more high performance animal and can't tolerate big fluctuations in diet and care. As a result, she's much more demanding of a routine. If you fluctuate moving your beef animals by as much as two hours a day, they can tolerate it. But if you're two hours late getting to the milk cow, she's uncomfortable and stressed. It's hard on her udder, as well.

Fortunately, a family milk cow is not a commercial performance animal. If you give good hay in the winter, no grain, she'll reduce her production substantially, but she'll still give you way more than a goat. Unless she needs to go dry. Since the milk cow needs a couple of months' rest before having another calf, one cow can never keep you in milk year-round. That brings up one of the biggest questions on the homestead: do we keep two cows or do we go two months without milk?

Fortunately, today we can freeze extra milk and that question is not as acute as it was on early homesteads. The secret to freezing milk is to make it freeze fast. If it can freeze before the milk and cream separate, it'll be less freeze-tainted when you thaw it later. That's a question you can wrestle with after enjoying your first cow for a few months.

Milk cows are not cheap if you get them from people who sell to the family cow trade. I suggest you contact a dairy, preferably one that doesn't milk Holsteins (too big, too fragile) and see if you can get a tail-ender. That's a cow that barely produces enough to stay in the commercial herd. A cow they don't want is one they're willing to sell cheaper. You never want to buy a cow the owner wants to keep. Get the one the owner would like to cull.

Today, with all the herd share raw milk dairies around the country, availability is certainly as good as it was before many smaller dairies went out of business. Back in the day, a 10-cow dairy existed on every corner. That's not the case today, but the poor quality of industrial milk is definitely fueling a come back for the small grass-based dairy. Any dairy milking fewer than 150 cows would be considered a small operation; that's the kind you want to find for a tail-ender.

Don't be afraid to take a mentor with you. If you've never bought a milk cow, don't run into the purchase by yourself. All farmers aren't as honest as you might think. Don't buy one more than 10 years old. Something 4-6 years old is fine. A structurally sound milk cow should last about 14 years. You want a cow with some life left in her.

To be sure, since you're not pushing the production envelope, a cow showing some stress will no doubt straighten up when she's top dog at the bunk and gets your caressing individualized attention. A milk cow needs good forage and hay. If any animal responds to good controlled grazing, it's the milk cow. That fresh break of grass each day is her permission to give abundantly.

Traditionally, everyone milked twice a day. But more and more grass based operations that don't push their cows as much adopt a once-a-day milking. That's a game changer because it cuts your chores in half. In general, by the time you get your accessories, tie up the cow, milk, and clean up, you have half an hour invested per milking. Twice a day is a whole hour. That's a lot of time to devote to a daily chore. It's also a lot to ask a farm-sitter if you need to be away.

But once a day can be loaded onto the front of the day before the normal work day starts. Most once-a-day milkers prefer early morning because it leaves the evening flexible for work or socializing. If you start your day with milking, it's behind you early in the day and you don't have to think about breaking up your day. This schedule drops milk production by about 20 percent, but it drops labor by 50 percent. The tradeoff works. Besides, generally a cow gives you more milk than you need anyway. Not too many families drink a gallon a day. Although with good quality milk, you can guzzle it like water.

Let's talk logistics. Cows walk slow. Really slow. Milk cows walk even slower. Except in the winter when the cows were in the barn on hay, we always milked in the field. If you have to go to the field to bring the cow in, then milk her, then take her back out, even if the field is not far from the milking stanchion, it adds 20 minutes to the chore. Start timing yourself from the time you walk out to the field and get the cow into the stanchion. Then time the trip back out to the field and back to the house. You'll be hard pressed to do those segments in less than 5 minutes; four segments is 20 minutes. That almost doubled your chore time.

To save time and save unnecessary exercise for the cow (if she burns up calories walking, she doesn't have as many to make milk) figure out how to milk in the field. Here are some tips:

1. Make a simple shade mobile for sun and weather protection. Although shade cloth won't keep the rain off you, it'll deflect it a little. If it rains, don rain gear and don't complain. If you make an impermeable roof on your shade mobile, you'll have to make it extra heavy so it won't blow over. That's not worth it. Rain gear is a lot simpler and cheaper.
2. Keep some sort of pan at the shade mobile that you can put a couple handfuls of sweet feed in to attract the cow to come to you. Not much enticement is necessary to generate a cow's interest.
3. When she comes to the pan with some sweet feed, tie her with a rope or chain you keep there for the purpose. You could install a stanchion, but we always got along great with a simple piece of rope or a chain with a clip. You just want her to know she's not free. Anything that rubs on her neck will let her know she needs to stay put.
4. Keep something to sit on in the shade mobile. That's for you to sit on while you milk. Unless you're into squatting, a stool is far more comfortable.
5. Depending on how meticulous you want to be, bring everything else you might need in a tote. Of course, bring the bucket too. Your tote might have a brush, a squirt bottle of soapy water (I don't recommend anti-microbials) a clean cloth (paper towels are fine, but that's

something you have to keep from blowing around the field when you're done), teat dip if you use it, and bag balm.

6. Keep a couple pint bottles of frozen water in your freezer. Take them out to the field with you in the milk bucket. These little ice bottles will help cool down the milk quickly in the field. We sometimes walked more than a quarter mile to the field where the cows were to milk them. On a hot day, that walk time can reduce shelf life substantially. This is another reason to do the milking early in the morning, the coolest part of the day.
7. If you have other animals with the milk cow, they may smell the sweet feed and want to come over to eat it. Keep a stick handy to thwart efforts by others to pinch the cow's treat. After a couple of pokes, they'll get the idea.
8. Always milk a cow from her right side. Teach her to stand with her left leg forward and her right leg back to give you room to place the bucket and work. Handle her udder like you would a woman's breasts. Be gentle, no sudden movements. She's letting you have an intimate relationship; respect that. Don't try to pull her teats off. Rough hands make jumpy cows.
9. If more than one person milks, make sure everyone uses the same protocol. If you milk both fronts and then both rear teats, do that every time. Don't alternate it. If you milk diagonally (that's the way I like to milk) start with the same pair every time. Milk from the same side. Make sure every milker knows the procedure to not upset the cow. Remember, she only lives in the moment.
10. Don't dilly dally once you've milked. Release the cow and get up and away to the strainer and refrigerator. All dairy activities revolve around temperature.

I can't imagine any reason to get a mechanical milker until you're milking more than four cows. At that point, though, you're running a dairy. The harsh cleansers and risk of bacteria in the lines coupled with the time to flush and clean make the machine uncompetitive until you're doing several cows. A simple milk bucket and strainer, too, does not stress the milk with pumps and pressure.

I'm a big believer in natural service. You can use artificial insemination, but unless you keep a semen tank and know how to do it yourself, calling a technician is risky. It has to be done at just the right time or you'll miss the window of opportunity. If you have a good relationship with a neighbor, offer to keep his bull for a month. You'll know when the cow is in heat when the bull is there; when he serves her, take the bull back.

I know everyone reading this who wants a milk cow will try artificial insemination. That's fine; get it out of your system. The reason is because everyone hopes for a heifer calf. A lot of dots have to line up to use artificial insemination, get it to take, have a heifer calf, and have a calf that works for you. It's a leap, but you won't know the wisdom of my suggestion until you try. Ha! You might be fortunate, and I hope you are. But it's a stretch.

Finding a dairy bull for natural service, near enough to your homestead, is nearly impossible. You'll have far more success using a beef breed bull. That means you'll be buying a new cow about every 6-8 years. Even a smallish dairy culls several cows a year. If you can find an outfit that's not into every techno-glitzy industrial procedure, chances are one of these lower producing cows will work for you. Often the reason a cow is a lower producer is because her genetics are less hyped than the top producing cows; that's exactly what you want for a family milk cow. The dairy farmer is glad to get a couple hundred dollars more for one of his cull cows than he would if he took her to the sale barn.

Buying the cow and using beef bulls to service her solves all the AI fragility and risk. Usually a beef cattle operation in the area will have an extra bull. On our farm, we keep a couple of extra ones in case one goes down during the breeding season. Because you're milking the cow when you want to breed her, you can't take the cow to the bull; you need to bring the bull to the cow. If the cow is healthy and cycling, you should only need to host the bull for a month.

Ideally, you'd keep him two months just to make sure. Often, a farmer is happy to let you feed his bull in exchange for the service. You might need to check around with several farmers before you find one willing and happy to work with you on a trade type of arrangement. In some cases, you may need to go an hour away to find the bull. This is an important decision; if it doesn't feel quite right, move on until you find a bull and owner that fits your needs.

If you can't find a bull and are forced into using AI, use a bull with low birth weights and preferably one that has some age on him. If he was a problem bull, the AI company would have pulled him from the pool. If he's more than six years old, he has lots of calves on the ground and his track record is well established.

The best way to know when a cow is in heat is to have a steer with her. If you're new to this, don't trust yourself to be able to tell when she's in estrous. Some cows are more demonstrative than others. Fortunately, dairy cows are more demonstrative than beef cows; that's in your favor. If you use a beef bull and get a heifer calf from your dairy cow, that calf can be an excellent family milk cow. She'll never produce as much as her mother, but she'll produce enough and be easier to keep.

Now let's discuss the calf. To nurse or not to nurse, that is the question. It's a huge question. What do you do with the calf when she births? Do you raise it on the bottle or do you let it nurse, and if so, how do you continue to get milk? Lots of questions. A milk cow tries to meet her volume challenge. In fact, in commercial dairies, they practice what is known as challenge feeding. Cows that have the potential to produce more milk receive more concentrates to challenge them to produce more. Transponders hanging around the cow's neck activates a feeder when she pokes her head in and based on performance, it releases the correct amount of feed to match with her production.

Your family milk cow wants to supply all that's demanded of her. That's why you always want to strip her out clean when you milk. If you leave some milk in the udder, she'll assume she doesn't have to make as much and reduce her production. By the same token, when she's constantly milked out, she does her best to keep up with the demand. This can be detrimental to her health if she stays empty all the time.

Desperate to satisfy the demand, she will draw from her own reserves to try to keep that udder performing. I'm belaboring this a bit to explain the danger in nursing. A cow with milking genetics will always produce way more than a calf can drink--at least at first. If you milk the cow and leave the calf on her, the two demands can easily pull her down. She'll get thin and unthrifty and might not rebreed.

Remember, once she calves, she needs to build back her body's reserves while she's milking in order to ovulate and rebreed within about two months of calving or you won't have a calf next year. If she's a cross between a milk cow and a beef cow, like many family cows are, this is not quite as critical because her beef genetics preclude her from being as productive as a purebred dairy cow. My favorite milk cow, Whitey, you might remember, was half Angus and half Holstein. Her nickname was Beef because she sure didn't look like a milk cow. But she was a sweetie. In her prime she would give 4 gallons a day.

Especially if you're a novice and unfamiliar with body condition, one of the easiest mistakes to make is letting your milk cow draw down too thin. It happens slowly. The bullet proof way to handle the calf is to let it nurse for about three days to clean out all the colostrum. That gets the calf healthy and lets the cow naturally bond for a bit, which helps her hormones adjust to normalize all her birthing adjustments. Then take the calf away and raise it on a bottle.

New Zealand vacuum nipples are much better for this than calf bottles you buy down at the farm store. The reason is two fold. First, when a calf nurses, it generates slobber that aids the digestion process. If you ever notice a nursing calf, you'll notice slobber actually dripping to the ground. If a calf nurses too quickly, like from a bottle with a worn nipple tip, it doesn't work at it long enough or hard enough to create that important slobber. The result is poor digestion. The vacuum nipples developed in New Zealand force the calf to nurse aggressively enough long enough to work up a slobber.

Secondly, a calf has four stomachs. The first stomach is for roughage and the second one is for cud. In other words, when a cow is grazing and swallowing fresh grass, it goes in stomach number one. Later, when she coughs it up and re-chews it, she swallows it into stomach two. If you watch closely, you'll see a cow swallowing cud kind of kink her neck. She's directing where that cud goes. For good digestion, milk should not go into a calf's first stomach. It should go into stomach number two. To make that happen, it needs to kink its neck, like it does when it's nursing.

A cow's udder is always too low for a calf to stand and nurse with its neck out straight. No, it has to kink its neck to get down low enough to nurse. That's the position we want a bottle fed calf to be in when it's nursing. Again, the New Zealand vacuum nipples are low enough to make sure the crooked neck happens.

Don't over feed the calf. It needs milk at least two times a day and preferably three at least for the first couple of weeks. Once it starts picking at some grass and growing, twice a day is fine and never more than a gallon up until one month old, then you could increase it to a gallon and a half. Like I said, this is the easiest and probably most foolproof way to handle the calf.

After three months, you can decrease to a gallon a day and after four months, half a gallon a day. Sometime in the fifth month, you can quit altogether.

Here are a couple of other options, both generally termed calf sharing. Both require your attention twice a day. In the first scenario, you let the calf run with the cow all day and pen it up at night. Milk her in the morning and put them back together after milking. The obvious logistical problem with this is putting the calf in a tight enough pen to hold it until morning. The calf grows bigger every day. The calf is also not stupid and soon learns this routine. The calf begins getting cagey and harder and harder to separate from mama in the afternoon.

It also gets more wily about getting back with mama overnight and might keep you up once in awhile voicing unhappiness. Both cow and calf must be kept in secure enough places to keep them apart. If they break out and get together a couple of times, you're in big trouble because you have animals going rogue on you. If both cow and calf are out in the field all day, how do you get them separated into secure areas for the overnight separation, especially when both want to stay together? That's a logistical problem. The temptation is to keep both cow and calf in the barn where separate pens can be maintained and you aren't chasing them all over creation. Of course, that compromises all the grazing goals of the homestead.

As a result, the second option is to put a halter on the calf and tie a 50-foot piece of rope to it. Remember to adjust the halter as the calf grows so it doesn't get too tight. The calf drags this rope around easily all day but in the evening, you can go out, grab the rope, and lead the calf to a secure corral or pen. Again, the calf is not stupid and will become wise to this trick too. But the long rope lets you grab the calf without getting close enough to frighten it (enter its flight zone).

While this all sounds simple, it can become quite a rodeo, especially once the calf gets big. When the calf hits about 300 pounds, it's going to be a handful. You might have to call out reinforcements. But like all things routine, normally both the calf and cow acclimate to the procedure and it becomes doable. As the calf grows, of course, you can start withdrawing it for longer periods of time and then finally wean it altogether.

Electric netting was not invented when we quit milking our family cow. Homestead guru Justin Rhodes uses this marvelous invention in a new wrinkle. He does a leader-follower program with two net circles using the cow and sheep. The milk cow is in the first circle and the sheep in the following circle. He puts the calf in the sheep net overnight and lets it in with its mother after milking in the morning. The calf offers some predator protection for the sheep and the cow and calf can see each other at all times, reducing stress. Justin agrees that routine is everything, and both cow and calf acclimate to this disciplined procedure, primarily because they are always proximate to each other.

About the time you wean the calf, the cow's milk production starts to drop off a little and you'll be ready not to share it anymore with a calf two or three times bigger and stronger than you. Weaning nose rings can work as well once you actually want to wean. This is a simple clamp you put in the calf's nose, with a spiked arc, that pricks the cow when the calf wants to nurse. The pricked cow kicks and walks away, denying the calf a chance to nurse.

While these nursing options sound a bit difficult, they offer one big payoff: if you want to go away for a few days, nobody has to milk the cow. You can leave the calf with the cow and know that all is well. These flexibility options make big differences on homesteads. A commitment to animals is not a light commitment. Plants don't run away and don't need daily attention. Animals are a different story. As uneasy as you might feel about the nursing idea, it does offer a freeing option for your family.

Few animals bring more joy than a milk cow. I'll leave all the goodies you can make with the milk and cream to other books. I often say that my favorite way to drink milk is with a spoon in ice cream. For a myriad of reasons, a milk cow complements the homestead's repertoire and eclectic self-reliant fare.

Chapter 24

Rabbits

Everything in this chapter I've gleaned from my son, Daniel, who proofread all of this for accuracy. He's been raising rabbits now for more than 30 years, starting with a pair of does and a buck when he was 8 years old.

My older brother had rabbits when we were teenagers, giving me early exposure to rearing and chasing bunnies. His mentor was an elderly man about 15 miles north of us who was one of those kind of rural eccentrics. He loved rabbits and I loved listening to his stories and watching his eyes sparkle when talking to my brother about how to raise them. Few things make an old man dance like a young man interested in what the old man's doing.

In fact, rabbits accounted for our family's earliest forays into mobile infrastructure. For years Dad worked on plans for mobile shade big enough to accommodate cows and eventually built it. But for rabbits, the prototype only took a couple of weekends. One of the greatest advantages of small animals is that their infrastructure is small enough to be built easier than things for cows. Downsizing is a real advantage: rabbits and chickens offer that opportunity.

Dad designed and built a mobile rabbit system with a central hutch divided into four quadrants for four does. At 6 ft. X 8 ft., each hutch quadrant was 3 ft. X 4 ft.. A mesh floor allowed all the manure to exit easily. This 2 ft. high hutch acted like the central portion of an airplane. Two wings 8 ft. X 12 ft. and divided in half lengthways gave each doe a 4 ft. X 12 ft. run of pasture. Every Saturday, when he was home from his town accounting job, we would go out and move the three pieces: hutch core and two wings.

With Dad on one end and my brother and I on the other, the three of us could pick it up and move it forward 8 feet to a new spot. We encountered difficulties almost immediately but didn't give up for months. Rabbits dig. Little rabbits don't need much of a hole to escape. Any little depression in the ground or digging exercise by the rabbits yielded an escape opportunity. Since rabbits are nocturnal, their overnight progress and escapades greeted us on many mornings.

We realized early on that we needed to do something on those wing runs to keep the rabbits contained. Another rabbit distinctive is that they aren't easy to catch. You can run down a chicken and catch it. But a rabbit? Forget it. A rabbit out is a rabbit at large. The first idea was to put finishing nails every inch into the bottom board of the runs, and clip off the heads with nippers to sharpen them. Of course, that made it impossible to slide the run, but the three of us could pick it straight up and set it down for the move.

The nails didn't work. Perhaps if we'd had a manicured croquet field, it would have. But on a farm pasture, slight undulations are everywhere. The next idea was to put a 12 inch poultry wire fringe around the inside of the perimeter. That way they would have to dig longitudinally to get out. That helped, but still no dice. The main thing I remember about the whole pastured rabbit thing was sitting with my brother, hiding behind a barn corner or a tractor wheel, long twine in hand tied to a prop stick on a box with a carrot under it, waiting for Brer Rabbit to make his appearance.

We made nets, dug rabbits out from under stacks of things and generally spent hours trying to corral these elusive bunnies. Eventually Dad gave up on the mobile rabbit idea and we went to an inside hutch plan with cut-and-carry (bringing in fresh grass and greens like comfrey). Not ones to throw anything away, we pulled those now defunct rabbit runs up into the rafters of the barn--"we might find another use for them some day"--and went on about other things. A few years later when my laying hen project outgrew the backyard chicken coop, Dad suggested that we bring those old rabbit runs down out of the rafters and put chickens in them. Folks, that is the story behind the development of our pastured chicken enterprise. The whole thing is exactly that serendipitous.

When my brother went off to college and shut down his rabbit business, we took a hiatus from raising rabbits until Daniel took an interest. A family at our church moved from the country to an apartment in town and needed a new home for their two does and buck. Daniel agreed to take them and we quickly built a little shelter for them and he ran them in the yard. Being a tiny shelter, in the yard, and mature rabbits, we didn't have any trouble keeping them in.

Rabbit meat is certainly not a mainstay, but it's versatile. When anyone asks how to cook a rabbit, we just respond that you can cook it any way you would cook a chicken. The pieces are of similar size and the young rabbits are called fryers, like a very young chicken. Rabbit tends to be lean because it stores fat in its viscera, not in its muscles. This gives it the distinction of the most concentrated protein of all the domestic livestock. Because of the protein punch it packs, it satiates hunger with the least number of ounces.

Rabbits are unique in that they eat like an herbivore but their digestive system is monogastric, like a pig or human. This seems impossible because monogastrics can't survive on grass and twigs. Enter the magic of cecotrophy. This is cool. When a rabbit eats, the material enters its stomach and then separates into digestible and indigestible components. The coarse fiber components go out the colon and excrete in little brown balls we all know as gorgeous rabbit poop, or pellets.

The digestible material goes into the small intestine and then into the cecum, which is a large elongated fermentation tube. You see, digesting cellulitic material like grass and twigs requires fermentation. Herbivores accomplish that process in their first stomach, the rumen. Rabbits do it toward the end of their process in the cecum. This large organ creates little soft shiny balls called cecotrophs that enter the colon and exit. The rabbit bends around, grabs these balls out of its anus, and re-eats them. Some material goes through this process four times as the rabbit extracts every bit of nourishment.

This process is why in some religious circles people debate whether rabbit is clean or unclean (according to the Levitical law in the Old Testament). Some argue that this process is really a sophisticated type of cud chewing, which makes the rabbit clean. Others argue that it's not actually chewing since it's being done down in the digestive tract. I won't answer that today; I even eat pork, so I'm not a good judge.

The rabbit's efficiency is legendary. Think about a cow that produces a calf each year. If the cow weighs 1,000 pounds and the calf 500, that's a total offspring production ratio of 50 percent (500 pound calf divided by 1,000 pound mother). One doe that produces 5 litters of 7 kits each takes her 9 pounds and converts it into 158 pounds, assuming a fryer live weight of 4.5 pounds. Let's see, that's an efficiency of 1,755 percent (158 pound kits divided by 9 pound doe). If that doesn't get your attention, I don't know what will.

Rabbits may be the most underrated of all the homestead livestock options. And if you happen to live in the city, it's by far and away the best option because rabbits are quiet, unlike chickens. In France, the per capita consumption of rabbit is about

14 pounds. If Americans ate rabbits like that, we would need 4.62 BILLION pounds per year. If any meat option is conducive to small spaces, backyards, and homesteads, it's the amazing rabbit.

Unfortunately, American culture places the rabbit emotionally off limits with everything from folklore's iconic Uncle Remus stories hero Brer Rabbit to the Easter Bunny and Thumper in Bambi and of course Mr. McGregor's nemesis, Peter. Trust me, when you start raising rabbits for meat, all of these famous rabbits become part of the banter around the butchering, cooking, and dining processes.

The chef who most influenced and enabled our early foray into vending to restaurants had a wonderful sense of adventure and humor. One spring she decided it would be hilarious to put rabbit on the menu for Easter Sunday. For some strange reason, her patrons didn't appreciate it. I think that went down as one of her biggest chef blunders. She told me later, with a rueful grin: "I learned there's one weekend you'd better not put rabbit on the menu: Easter."

If you've read this far in the book, I'm assuming you're not a militant "my cat is my aunt is my chicken is my niece is my dog" person so let's get right into the nitty gritty of rabbit production. I've already told you about Daniel's early bouts with diarrhea as he moved his rabbits to forage based genetics. He brought good hay, grass, and especially comfrey (it's especially nutritious and easy to cut and carry) to the rabbits. Based on my childhood experiences with rabbits, we were extremely gun shy about raising rabbits on pasture.

Rabbits are exceptionally prone to respiratory problems like *pasteurellosis* and to an insidious protozoa called *coccidiosis*. If you think about it, in the wild, rabbits don't cluster tightly. Even in a native rabbit warren, they each make their own holes and though they live in community, each nest is quite separated from the others. This separation is part of the protective barrier against respiratory and infectious diseases.

In a housing situation, chickens are fine with four changes of air a day while rabbits should have eight or more. Fortunately, chickens don't mind more changes of air. While we were aware of numerous attempts at colony raising and pastured communal models, for hygiene, population and inbreeding control we opted for a caged system. Rabbits are quite territorial and can kill each other fighting, eat each others' babies, and are highly susceptible to *coccidiosis* when in close contact with the ground. If you want a more exhaustive look at this issue, I highly recommend Nichki Carangelo's wonderful book *Raising Pastured Rabbits for Meat: An All-Natural, Humane, and Profitable Approach to Production on a Small Scale*. She dives into the weaknesses of these seemingly more natural systems and pretty much arrived where we arrived: caged does and pastured grow-out.

One of the strange defensive mechanisms of does, when feeling threatened, is to eat their bunnies. In colonies and outside, the stress risk factor is higher. In a colony you have bully does. Out on pasture, things go bump in the night. Fortunately, rabbits are nocturnal, making them quite aware of any predator that comes around. Unlike a sleeping chicken, the rabbit sees the threat and moves to the middle of its shelter, away from the edge. But if a predator is circling, or if a visitor's child screams, the doe can be terrified and eat her bunnies. An undisciplined farm dog can play havoc with a doe and her new bunnies. For all these reasons, a more protected environment is desirable.

In general, you have three options for dealing with the manure and urine. One is to clean out under the cages every week. If you don't, the rabbitry will stink and worse, the ammonia vaporizing from the urine rises up to the rabbits and encourages weepy eyes and respiratory problems. The second is to put worm beds underneath and turn everything into castings. That requires steady carbon additions to mix with the manure and urine. This works as long as you can keep the worms well above freezing. If your quarters aren't cold-proof, the worms will vacate when things get too cold. No fun.

The third option which we stole directly from the stacking concepts in permaculture, was to use a deep bedding system with chickens as the stirring machine. This sheet compost under the rabbits, managed by chickens, developed into what we call the Raken House (rabbit + chicken = raken). Again, the plans and specs for this are available in *Polyface Designs*. The carbonaceous deep bedding absorbs all the urine and the chickens keep it aerated. This double-level system uses cubic space and squeezes a lot of production out of a small area.

To keep things composting and aerated under the rabbits really only requires one chicken per rabbit. The square footage, however, can handle more chickens than that so we put in whatever the floor space can handle. The chicken floor space can include your service alleys. About one 5-gallon bucket worth of carbon every two weeks per rabbit is enough to maintain the deep bedding system. In our more commercial set-up, we simply start with 18 inches of carbon and don't add carbon as an ongoing chore. With the chickens in there, adding carbon efficiently is problematic because it scares the chickens--too much activity.

Let's get practical, utilizing something the size of a two-car garage, perhaps 20 feet X 20 feet. That's a total of 400 square feet. If each rabbit hutch (we call them holes) is 2.5 ft. X 2.5 ft., that's 6.25 square feet per rabbit. Of course, you can't cover every square foot because you need room to walk. Let's start with the perimeter of 80 feet. We'll leave 10 feet for a nice big door access so we can get in with a tractor front end loader or wheelbarrows easily. That drops us to 70 feet which is enough room for 28 holes (70 divided by 2.5 linear feet per). Those perimeter holes occupy 175 square feet.

With 5 feet (2.5 + 2.5) perimeter lost, we still have 15 feet to work with in the center of the structure. That's plenty of room to put one set of holes down the middle and still leave 5 ft. between for access. If you put 4 holes back-to-back down the center, leaving enough room around both ends to maneuver, that's a total of 36 holes (28 + 8 = 36). If you assume half are full of grow-outs and half are full of breeding stock, including a couple of bucks, that's conservatively 15 does, each of which can produce 5 litters of 7 kits, which is 35 bunnies per year, for a total fryer production of 525 fryers per year (15 does X 35 fryers = 525).

If each of those fryers yields a 2.75 pound carcass, that's 1,444 pounds of meat from a two car garage. Things never go as perfectly as paper plans, so to be fair, let's drop it to 1,000 pounds. That's a lot of production from a two-car garage, don't you think? But wait! We have laying hens underneath. At 3 square feet per bird, we can house 130 laying hens on the floor. You don't need to save any maneuvering space because they'll move out of the way as you walk around. Those 130 hens will lay a rolling year-round average of 50 percent, yielding 65 eggs a day. Let's be conservative and assume 5 dozen. If a year has 365 days in it, at 5 dozen a day on average, that's a total annual production of 1,825 dozen.

Again, to be fair and conservative, we'll drop that to 1,800 dozen. Now let's think about value. The 1,000 pounds of rabbit is worth $8 a pound. That's $8,000. The 1,800 dozen eggs are worth $5.50 a dozen. That's $9,900. Are you tracking with me here? We're talking $17,900 from a two car garage. If you think that's outrageous, drop it, because we all know nothing works as well in real life as it does on paper. Okay, drop it to $15,000. Drop it all the way to $12,000 if you want. Folks, that's some serious opportunity in a small space. Oh, sorry, this is about homesteading, not about earning a living. I do apologize, but when I get into this stuff my old entrepreneurial persona takes over and I can't help it.

In his 1974 classic *Raising Small Livestock*, Jerome D. Belanger starts off the whole book with rabbits, claiming they're the most overlooked option out there. Amen, brother. The chickens underneath really change things. They keep it sweet smelling and composting. We make compost from lots of sources on our farm: cows, pigs, chickens. By far and away the best compost we made is the from the Raken house.

Before moving to pasturing, in full transparency, we actually do not use the Raken house for rabbits and chickens year-round--we used to. But once we began building hoop houses for winter laying hen housing, we realized how advantageous that would be for the rabbits. No more frozen water lines. Early on, when the temperature went to single digits Fahrenheit, Daniel couldn't keep the water lines from freezing so he had a stash of little crocks he'd use. He'd boil water on the wood stove and take it out a couple of times a day, pouring it into the crocks that were half full of ice.

The hot water melted most of the ice, which cooled the water and made it lukewarm, which the rabbits loved. In the hoop houses, the water always thaws once the sun comes up, making the crocks and boiling water an ancient memory. One of our rites of passage in spring and fall, these days, is moving the whole rabbit breeding operation from the Raken house to the hoop house for winter, then moving them back in the spring. We've become quite efficient at this and it only takes a few hours. That opens up the Raken house for about 100 days each winter, freeing it up to house pigs. The pigs do the final turning of all the deep bedding. When the pigs come out in the spring, we clean it all out with a front end loader, put in fresh wood chips about 18 inches deep, and restart the process with rabbits and chickens.

This housing transfer now offers the decided advantage of host-free status for a significant portion of the year, which reduces pathogens. Resting the housing from its dominant species keeps everything more hygienic.

We feed an unmedicated commercial rabbit pellet as the TMR (Total Mixed Ration), or concentrate component, and then supplement as much as possible with everything from comfrey to hay to grass and with the weaned kits, pasture. Does are like sows; you have to limit feed them or they get too fat and won't breed. Gestation is 31 days and does are in estrous about 20 days out of 30. Always bring the doe to the buck, never the buck to the doe. If you bring the buck to the doe, he's more interested in sniffing around the new territory than paying attention to the doe. No distractions, please.

By bringing the doe to the buck, he's full-on focused on his business. If she's in estrous, she accepts his advances, he hops on, and dramatically flops off. This is sex education, farming style. When we do our seminars, this is always one of the most memorable moments. No foreplay, and no coy discussions. It's one thing on the mind, the whole procedure taking about 30 seconds to a minute. You don't need to leave the doe in there for a day or even 5 minutes. It either happens or it doesn't.

You need at least one buck to ten does. For many years we brought in no outside genetics, but in recent years we've brought in some new bucks and it's been beneficial. The big no-no is breeding full brother and sister. As long as you never get closer than a half, you can rock along for a long time. Like any good selection and culling process, always select breeding stock that's healthy and thriving in your conditions.

With a minimal 3 does and 1 buck program, you should replace the buck with outside stock every two years. A buck can breed his daughter but that's as close as you should let it get. Since a buck only lasts two years, you'll need a replacement buck after two years but you can't select from any of the saved does from the same litter or the relationship will be too close.

Because *coccidiosis* is highly prevalent in rabbits, if we want to save breeding stock out of a certain litter, we process the opposite sex siblings than the one(s) we want to save. If we want to save a buck, we process the female kits first and look for telltale signs of *coccidiosis* (white spots in the liver). You can be pretty sure that the brothers' insides will look about like the sisters'. If everything is clean, we keep a buck; if not, we process the whole litter. If we want to save a doe, we invert this procedure.

One of the great assets of rabbits is the luxury of offering this kind of look into the inside of close relatives before selecting breeding stock. I've often wished we could do the same with cows. The inside look is the best test and has been the backbone of our selection process for decades.

When weaning, we always take the doe away from the bunnies, not the bunnies away from the doe. The reason is similar to the breeding protocol. The bunnies, already a bit stressed from losing mama, don't need the added stress of relocating. They stay in familiar surroundings and the only thing missing is mama. Again, in our human mind, we see each 5 ft. square compartment as identical to all the others. What's new or different about shifting from this hole to that hole?

But animals are fully aware of the difference. Each compartment has a different view, different air movement, and different sounds. What seems identical to us is highly different to a rabbit. Moving the doe and leaving the bunnies behind enables them to get through the weaning process easier.

Little metal squares with numbers painted on and fastened with a J-hook move with the doe to her new hole. This way you can track both doe and bunnies as they move around. That's easier than stationary numbers on the holes and a spreadsheet tracking the moves.

Having done the exercise above, you can see that even a three-doe, one buck set-up can generate a couple rabbit fryers a week for your family. This square footage can be tucked in corners, under awnings, into solariums in tiny spaces.

Even at this rudimentary production level, you'll want to keep some records in order to keep up with the breeding, kindling, and weaning. Information you want to capture includes the following:

Breed date
Due date
Birth date

Number kindled
Number weaned
Buck used
Doe used

You'll want to track this information through the bunnies so you can look back at performance and genetic strength. Does usually don't drop off a cliff in performance; they gradually drop litter sizes. If you have a doe that's been a consistent 8-bunny kindle and then goes to 5 and then to 4, you can be pretty sure she's at the end of her productive life. Record keeping is the way to know what your trend lines are for each doe and how to select your new breeding stock.

Domestic meat rabbits won't perform well without access to feed concentrate, but that doesn't mean you should deny them as much natural plant material as possible. Don't worry, the rabbit is smart enough to know how much feed it needs versus how much roughage it needs. A favorite is twigs. You know how rabbits gnaw on small sapling trunks and bushes in the winter--in fact, wrapping young orchard plantings in protective material is largely to keep rabbit gnawing at bay until the tree trunk is bigger. In the spring when we prune orchard trees and grape vines, we put a couple of these prunings in the cages each day. The rabbits love them for their nutrition and it gives them something interesting to do all day.

You can source edible vegetative material from many places. Of course, if you have a yard you can use grass clippings. Don't use clippings after they've gone through a rotary lawnmower because the pulverized blades aren't palatable. If you have an old-fashioned reel mower, those clippings work fine. Snippers and pruners work fine; anything that doesn't masticate and pulverize the blades is acceptable. Any leafy greens you grow or acquire secondhand add delectable variety to the rabbits' diet. Of course, carrots, cucumbers and root crops in general are a great supplement. Any woody twigs, lawn trimmings, and hedge prunings bring in minerals and lots of good gnawing time.

Cage construction, water systems, feeding systems and nest boxes are all covered well in Nichki Carangelo's book *Raising Pastured Rabbits for Meat* and *Polyface Designs* so I won't repeat all that minutiae here. The only thing I'll say about that is to get the good stuff. Spend the money for the double dipped wire and the baby saver bottom and the side-mounted feeders and nipple waterers. You really do get what you pay for. We like 1X2 inch wire for the sides and top but 1/2X1 for the bottom, all fastened together with J-clips.

One of our former apprentices built a nice 10-hole system on skids that he pulls around his yard. During the non-freeze season, it works great. In the winter, he simply tows it into his hoop house. You always want the cages to be at handy accessible level, not low enough to require stooping to service. His arrangement is essentially two stringers legged down to a pair of skis. An additional refinement would be to make such a contraption match the size of your garden beds and simply move it from bed to bed as part of your fertility program. If it's too heavy or cumbersome to move by yourself, have a potluck (you supply the rabbit stew) and all your friends can grab a corner to pick up the assembly and move it to another bed. This is what community is all about.

Now to pasture. Once the Raken house proved its functionality, our thoughts turned to getting the weaned bunnies on pasture. We wean at six weeks and move them outside at seven. That gives them a week to acclimate in their birth place without their mother before moving to a totally different places. Every change is stressful; spacing out stress is critical to maintain health.

Rabbits like to start on the end of a blade and gnaw it down toward the ground. We initially tried covering the bottom of a pasture shelter with 2 inch poultry wire, but the rabbits too often inadvertently bit a piece of wire and became more reluctant to graze. Further, the holes pressed over the grass blades, prohibiting the bunnies from being able to get on an end and gnaw it down. We realized that idea had serious flaws. They still grazed some, but they definitely seemed impeded.

Then we struck on the idea of laying out the poultry netting, letting the grass grow up through it, and moving the now floorless shelter over it. The imbedded wire would keep the bunnies from digging out but the grass would grow vertically and not be bent over when we moved the floorless shelter over top. That worked beautifully for a couple of years. We laid out a pad to handle an 8 ft. X 8 ft. four-quadrant shelter in about a 40-day rotation. The pad was 32 ft. wide X 40 ft. long, big enough for four passes of ten days each.

For the first couple of years, we really thought we had the answer. The rabbits stayed in; the grass blades stayed vertical; the rabbits did well. But the third year we had a couple bunnies get out. How could that be? As we checked things over, we couldn't find the poultry wire. Where did it go? We dug down about two inches and there it was, buried. Folks, the soil had uppened over the wire and the rabbits could dig out between the bottom edge of the shelter and the buried wire.

Well, bummer. Who would have thunk that in less than three years this extremely intensive pastured rabbit model would grow a couple inches of soil? It was amazing. The other thing that happened was that we noticed way more *coccidiosis*. It was apparently building up slowly; we weren't giving enough land rest and not running other types of livestock over the area. Too much of a single good thing.

Back to the drawing board. We decided to try the best of all worlds and make a small 3 ft. X 6 ft. X 2 ft. tall shelter with a slatted floor. The floor ribs, on the inside, allow the slats (1/2 in. X 1 in.) to be affixed to the underside. That keeps the bottom, the underside, smooth like a boat. The ribs, 24 inches apart, and slat space, 2 open inches between, allow 70 percent exposure to the grass and more importantly, allow the blades to spring back up straight for prime rabbit grazing. Voila! These small shelters were light enough that even a small child could move them. We could keep litters together for genetic tracking. The bunnies could eat all the forage they wanted and we could move these shelters anywhere on the farm. They weren't stuck in one area.

We could mow under the apple trees, around the outbuildings, and not come back to the same spot a second time in a year if we wanted to give that long a rest. These are the shelters we use today. Yes, the slats eventually rot, but because we move them everyday, not as fast as you might think. Wood rots a lot faster if it sits in one spot where the soil bacteria can surround it unmolested. If it's skidding across the grass every day to a new spot, it stays clean and shiny from the skid scraping. Further, the slats give the rabbits some protection from actually being in constant ground contact, which reduces contamination from *coccidiosis.*

When we're moving these shelters, the rabbits sit on the slats like water skis. The shelter requires no dolly for moving because it's light enough and slides easily on the slats. Are you ready for a little more figuring? The weaned kits go out a week after weaning. We rebreed her about a week before weaning, which gives her nearly a month without her bunnies before she rekindles. This roughly 9-week cycle is not nearly as aggressive as commercial rabbitries, but still produces five litters a year.

Commercial industrial outfits shoot for eight kindles a year and typically wear out their does in a year. Our more laid-back system enables an average doe life span of up to three years and helps keep things healthy. Pushing systems to their breaking point is not always the most efficient, especially biological systems.

An average 7-kit litter goes in an 18 square foot field shelter sometime during the seventh week of life. Our goal is a nearly 3 lb. fryer at 12 weeks. Sometimes they grow as long as 14 weeks; a lot depends on the weather and the time of year. Rabbits definitely prefer spring to winter, and moderate temperatures to extreme hot or cold. Remember, rabbits like to live in the ground, where the temperature is always about 50 degrees F. Without a doubt rabbits prefer cold to hot.

If these 7 bunnies get 18 square feet a day, in 5 weeks of pasturing before processing that's 35 move spots at 18 square feet per day, or 630 square feet. Our experience indicates well-bred pasture-genetic rabbits will eat well over half of their feed from the forage. This saves about half the feed costs. But let's not digress. These 7 rabbits will yield about 20 pounds of carcass, which includes both the meat and bone. This is a yield of about 1,380 pounds per acre (43,560 sq. ft. per acre divided by 630 sq. ft. of grazing). At $8 per pound, that's a gross income of more than $11,000 per acre (1,380 lbs. X $8 per lb.). That's just with one pass in a year.

Most places you could do two passes in a year, or even three. How much is an acre of grass worth? Conservatively around $20,000. That's not net, but it's a pretty decent return, especially when you realize you could graze this same acre with cows, run broilers across it, some turkeys, some layers. Oh goodness, here I go again talking like an entrepreneur when we're supposed to be talking about just feeding your family on a homestead. Somebody shut me up.

This whole exercise leads us to a significant point in the big picture. The smaller the animal, the greater the return per acre. The larger the animal, the less the return per acre. But here's the flip side of that coin. The smaller the animal, the more labor per acre; the larger the animal, the less labor per acre. In case you haven't figured it out, there ain't no free lunch.

The smaller your acreage, the more you want to concentrate on small animals. The smaller animals yield more per acre, but it takes a concomitant amount of management to tease out that high production.

Allan Nation, when making this point, would always say in an echelon of return per pasture acre the least was the cow, then the sheep, then pigs, then turkeys, then chickens, then rabbits, then guinea pigs, and if you really wanted a high return, you'd put electric fence half an inch above the ground and graze escargots at $150,000 per acre. Everyone would laugh but the point stuck. It's a good rule to remember.

Compared to all the other livestock, rabbit is certainly one of the easiest to process. It requires the least equipment for sure: just a couple of string loops hanging from a bar. For many years our clothesline T end served fine; eventually we moved it into the chicken processing shed. All you need is a loop to hang their feet apart. Our preferred killing technique is a swift strike to the back of the head to render the rabbit unconscious and then we hang it from one foot, slit the throat, then we hang the second foot.

When picking up a rabbit, grip it in front of the hind legs, back in your palm, rabbit hanging down. This leverages a pressure point in the back that keeps the rabbit from kicking and struggling. It's actually fairly relaxing. See the photo on page 87 for the proper way to hold a rabbit. No matter how good you get at handling rabbits, though, you'll get scratched occasionally. Anytime someone comes around who claims to have something to say about rabbits, Daniel asks him to roll up his sleeves and show off the battle scars. Kind of like a chef always has burn marks on her arms. It just happens.

The benchmark for processing should be under 5 minutes, start to finish. You should be able to do 12 an hour; Daniel can do 20. We either compost the guts or feed them to the laying hens as fresh animal protein. While that may sound gross to us, it's like ice cream to a chicken.

A rabbit sheds just like any other animal. In the summer, the hide is thin and not as furry. It thickens in the winter. If you're up on traditional tanning methods, you know that supposedly every animal has enough tannin in its brain to tan its hide. Brain tanning is doable, but a lengthy and involved process.

Daniel has experimented with all the different tanning methods but so far hasn't found one efficient enough to create a viable profitable enterprise. As a hobby, though, I'm convinced that tanning a few hides for mittens, gloves or hats would add more value and respect to the rabbit's life. We've given many hides away over the years to folks who want to try their hand at some home tanning.

Really the issue is as much political as it is process. Can you imagine the looks you'd receive if you walked through an airport wearing a rabbit fur jacket? Would it be beautiful? Yes. Would it attract views as a fashion statement? Yes. Are you ready to be bullied? Maybe no. The animal skin market is in the doldrums not because we've forgotten how to tan, but because we've forgotten how to connect to life. We've substituted petroleum-based synthetics and called it noble. What a shame.

Rabbits definitely have some fragility. They aren't bullet proof by any means. People who think you can throw a couple of rabbits together and stand back collecting bunnies have never tried to do it. One of the most challenging aspects of rabbits is the extensive record keeping required. A cow has one calf a year. A doe has up to 30. That's a huge difference in tracking, knowing the matings, the age, the kindling dates so you can get the nest boxes in on time. Good records will keep you efficient in your production.

I hope this exposure to rabbits gets you to thinking about possibilities you haven't thought of before. For a small homestead, rabbits offer self-reliance in an abundance hard to find anywhere else.

Chapter 25

Urban Possibilities

What if your land base is either urban or smaller than a homestead? Urban garden areas, both public and private, could certainly benefit from animals. Any commercial eatery could utilize chickens to handle food wastes. Here are some ideas for urban spaces where pasturing, even on a small scale, is not possible.

If you have a postage stamp yard, you don't want a mobile chicken shelter because you won't be able to enjoy your little grassy spot without stepping or sitting in poop. And your kids will track poop into the house every time they go outside. Even if you only have two chickens, you need something 4 ft. X 4 ft. That's 16 square feet.

If you move it every day on a 60-day rotation, that's 960 square feet. What does that look like? It looks like 32 ft. X 30 ft. In many urban areas, that would be a pretty decent yard, but as you can see, it's still too small for practical pasturing. If you have that little space, it should be in vegetable garden rather than chickens.

But the utility of chickens for the household is still a profound thing. Here is my suggestion if space prohibits pasturing or lawning the birds.

First, even though my rule of thumb for commercial housing is 3 sq. ft. per bird, you can't put two chickens in 6 square feet. Behavioral and social needs do not scale linearly from large flocks to exceptionally small flocks. At our farm, when we have 1,200 layers in a 3,600 square foot hoop house for winter housing, they don't spread out evenly. A clump gathers over here, another clump over there. They don't each stake out a 3 square foot spot and stand there. They gather and move and flow.

Because they are flock oriented, the birds enjoy being close to each other most of the time, but they also enjoy the option of space if they want to get away. Most of the chickens in the larger flock, then, occupy only a square foot. That gives the appearance of lots of open space. If half the birds only occupy 1 square foot because they want to commiserate, that opens up 5 or 6 square feet per bird for the rest of the flock. That looks downright spacious.

As you drop down to tiny numbers like two, the ability to gather and disperse within a larger context doesn't exist. Two birds in 6 square feet can never get very far from each other. They're always up against a side wall. That is why it's not a linear scale of 3 square feet per bird from 2 birds to infinity. To be happy, a hen doesn't want to feel constrained or hemmed in all the time. Something 4 ft. X 4 ft. is a minimal size, in my opinion, to allow the birds a modicum of flapping, jumping, stretching; just the minimal chicken stuff.

That size gives two birds 8 square feet apiece, which is completely unnecessary in larger flocks due to the spatial unevenness of the individuals. This is an important concept because if you don't increase space at micro numbers, the animals will be stressed all the time. Just for discussion, assume a pet Atlantic Canary in a bird cage in your house. Think about the size of that bird compared to a chicken. What would you say, maybe 1/5 the size? The cage is probably at least 2 feet in diameter, which is a bit more than 3 square feet for a 1/5 the size bird. At equivalency, that would be about 16 square feet for a chicken.

My point is my minimal 4 ft. X 4 ft for two chickens may sound big, but it's not really. It's adequate, but certainly not palatial. If you have more room, great. Just to close the loop on this discussion, what if you have a flock of 20 laying hens? At that point, you can get by with 4.5 square feet per bird; something 10 ft. X 10 ft. is plenty adequate. I've never seen a graph lining this principle out, but it's real. In general, I'd say the 3 square ft. per bird kicks in at something more than 50 birds. Obviously this is a subjective call, but I hope you get the idea. Perhaps some day we can put biometrics on chickens and measure stress levels at various spacing and various flock sizes.

My completely unscientific but experiential formula for stress is this: DENSITY X TIME X MASS = STRESS. You can put chickens in a crate at half a square foot per bird for a couple of hours and they're completely calm and content. But if you leave them there for a day, they get extremely stressed--that's time. A crate does not have very many birds in it; if you put 1,000 birds together at the same density, they would be incredibly stressed even for one second--that's mass. Finally, if you put just two birds in a tight space, they'll be stressed automatically--that's density.

Animals don't like to be scrunched up against a wall. Remember the flight zone. Cows never like to be pushed too close to an electric fence, for example. That's why we make all of our access lanes 16 feet wide. Chickens are the same way. Notice that all my smallest coops are 4 ft. X 4 ft., not 8 ft. X 2 ft. even though that is the same square footage. A long narrow alley like that stresses the chicken because she can't get away from the wall. She'll spend most of her day agitated, running back and forth and back and forth. Give her at least 4 feet from wall to wall to enable her to get to a middle and away from it.

If you're looking for my advice on micro flock coops size, here is a totally subjective list:

2 16 sq. ft.
3 22 sq. ft.
4 26 sq. ft.
5 30 sq. ft.
6 34 sq. ft.
7 38 sq. ft.
8 42 sq. ft.
9 45 sq. ft.
10 48 sq. ft.

Let's return to our two chickens in our 4 ft. X 4 ft. minimalistic coop. Offering 3 feet of head space and mounting the nest box on the outside saves space inside the living area. A nipple waterer and a side-mounted feeder also save space. Anything you can do to preserve the full living space by hanging essentials on the outside wall is helpful. Now we get to the fun part.

This coop is not going to be like anything you've ever seen in any micro-scale chicken rearing book. Rather than having the chickens on a wire mesh floor or newspaper, we're going to have these chickens on deep bedding. Are you surprised? Essentially, we have a deep box, preferably at least 2 feet deep, that can hold a bunch of carbon. We start these chickens at a one foot depth with whatever we can scrounge: some leaves, pine needles, sawdust, vacuum sweeper cleanings, biodegradable soft packing material. We'll put in a gallon or two of soil on the top, preferably with a few worms in it. The soil is a biological inoculant to ensure and facilitate decomposition.

Then we put the chickens in. If you can put some treats like kernels of small grain, popcorn, apple cores or squash seeds in there, all the better. You're trying to get the hens to attack this bedding with incentives to teach them that it contains goodies.

You won't watch any TV for awhile because you'll be mesmerized by these happy hens scratching, pecking, fluffing, digging--I mean they will tear that litter to pieces digging for gold.

If a portion of it caps on top, use a little trowel or gardening harrow to stir it and help the ladies. Of course, you can put all your kitchen scraps in and any weeds you pull from garden beds. If you have potted plants and you want to repot, throw the old potting material in there. They'll rip up roots, stir it in, and add it to the decomposition. Because chickens don't urinate, the material might need a little more moisture. If it does, throw in a couple glasses of water.

This whole process is as much an art as a science. If maintained properly, the coop will not stink; it'll smell like woodsy soil. You can add carbon along to keep the balance right. Remember that as it decomposes, the height drops. Never let it get below a foot deep, and slowly add material until it's 2 feet deep. When it gets to the top of your box, you may need to put some temporary board strips around to keep them from scratching too much material out. When it's full and the material a foot under the surface looks more like compost than distinguishable material, you'll want to clean it out.

When building the chicken coop part, which will undoubtedly be primarily poultry netting or some kind of mesh so you can see the birds and they can see you, you'll want to hinge a wall as an access panel. If it's longer than 6 feet, you'll probably want to break it into two hinged panels. You want to be able to reach into the box, stir the material if necessary, dump new stuff in, and of course eventually to scoop it all out. The deep bedding will never come up to the hinged door panel so don't worry about holding the material in there. It'll always be lower than the access panel.

Don't be surprised if the time from start to clean out is a year or more. Decomposition and worm castings often reduce mass of the original material by a factor of three or more. That means in essence you put in 6 feet of material, but it turned into only 2 feet. Isn't that cool? This will be the richest material you could ever want for your garden, houseplants or fruit trees.

Now, let's take this whole arrangement one step further. Imagine that if on top of the chickens you had rabbit hutches. Normally in tight spaces you'd want this contraption up against a fence or wall to minimize access area. I can imagine some instances where you might want it in a square rather than rectangle. That's fine. As long as you respect the overall square footage necessary for happiness, you'll get along fine. Here is the total height: 2 foot litter box, 3 foot chicken coop, 2 foot rabbit hutch. That's a total of 7 feet.

In a most extreme example, like a condominium where this set-up needs to be actually in your house or apartment, you can do that. Ceilings are usually 8 feet, so you have plenty of wiggle room. The main issue in such a case is dust. The rabbits won't make dust, but the chickens and all that scratching surely will. Ideally, you'd wall off the room and have an exterior window that you could open for fresh outside air.

If you are reading this nodding in understanding and contemplative acceptance, saying subconsciously "wow, this is really doable. Yes, I could do this in Manhattan" then you're my kind of kindred spirit. If you're laughing your head off that this nut actually thinks you're going to do this in your house, I can take it. But consider this. The average pet dog generates as much manure as 9 chickens. Not 2, not 3, not 4, but 9. Got that? And dog poop is nasty. Chicken poop is cool; I mean you could almost bake it into cookies.

Furthermore, what I've described is something 4 ft. X 4 ft. or a bit bigger. How much space does the average entertainment center occupy? Those massive pieces of furniture are often 8 ft. long and 2 ft. deep--that's 16 square feet, the same footprint as our dream rabbit, chicken, compost triple decker.

Or consider an aquarium or terrarium, or even a dog kennel, for that matter. If you have a dog that weighs more than 60 pounds, how big is the dog kennel where he stays from time to time? I guarantee you it's at least 3 ft. X 3 ft., which is not far away from 4 ft. X 4 ft. My point is that when you really think about it, we have plenty of things in our houses bigger than this. True, you might not be able to get something 4 feet wide through a door, but that's why you make this thing in three pieces. That way you can tip any piece sideways and bring it in.

Just imagine the entertainment value of a couple of chickens and a couple of rabbits in your condo. My goodness, you could charge folks to come and watch you breed rabbits. Better than Netflix, let me tell you. Let your niece come and gather a fresh egg out of the nest box. She'll go nuts and want to come see Auntie the chicken lady every week. You'll be the ultimate greenie eccentric practical authentic tree hugging environmentalist. Anybody can write a check to Nature Conservancy. Only the fearless, the courageous, the authentic would raise rabbits, chickens, and compost in their apartment. Go ahead, dare to be different. I dare you.

Remember when I talked about being unable to bring restaurant kitchen scraps back to the farm efficiently? Now I'll let the next shoe drop. What every restaurant needs is a small chicken coop by its back door. If you adapt the ideas here and move on up to 20 chickens, they will eat a pile of kitchen scraps. Putting these ladies in your dining room might be a bit over the top. Actually, I think it would be the coolest thing in the world. Can you imagine how folks would talk? Probably the health department inspector wouldn't like it. Just a hunch. But wow, that would make the coolest dining room in the city. Talk about creating buzz.

How about school dining services? Most schools have a campus with room to tuck a chicken coop behind the kitchen. It might need to be 100 chickens instead of half a dozen, but that's okay. In a day when schools don't bat an eye to spend $1 million on a gym, isn't it about time we spent a smidgen on closing our waste streams? The real beauty here is not just closing the waste stream, but in producing eggs. This is the best two-fer in the world. Now the kitchen doesn't have to buy eggs riding in on diesel fuel. Their waste doesn't have to ride out on diesel fuel. We cut the trucking both directions.

Dining services supervisors and chefs getting plaques from environmental organizations for adopting a composting program don't hold a candle to what I'm describing here. In the greenie-recognized composting programs, it's still a haul the eggs in, haul the scraps away--a segregated linear system all riding on diesel, trucks, and highways. In my proposal, we make a circle, eliminate the trucks and diesel, and reduce road congestion.

Folks, to me these are such simple, simple solutions that it makes my head explode to see academics and environmental think tanks trying to figure out how to stop global warming with fake lab meat. The tragedy of the human experience is that we make the simple complicated and the complicated too simple.

As nicely as this simple set-up fits with personal backyard gardens, it fits equally well with municipal and community gardens. Goodness, the Brooklyn Grange rooftop gardens on the naval yard have chickens on the roof. Surely any community garden could put in a few chickens to eat garden scraps, rotten vegetables, and in turn provide luscious eggs. What's not to love?

You see, dear folks, the reason people resist all of these ideas is because farm animals are universally considered to be smelly, noxious, and filthy. The tragedy is that too many, if not most of them, are. I won't debate that point. But it's like saying we shouldn't have any children since they're often obnoxious. No, they just need some love and discipline, some boundaries, some protection from violent video games and social media, and they're wonderful little beings to be around.

If you haven't figured it out by now, the whole well-managed, neighbor-friendly system depends on carbon. That's the missing link. I've seen more filthy stinky tiny dirt pooped-up chicken yards than you can imagine. Nest boxes full of poop. Filthy eggs. Soiled chickens. All it takes to revolutionize the situation is some diaper material. The carbon economy is both that simple and that profound.

Municipal landscape crews should be able to sell their carbon waste streams for millions of dollars. Fall leaves, chipped tree branches, ground up stumps--it's the secret sauce of micro-animal housing. Besides management intensive grazing, probably the single biggest positive feedback I've gotten in counseling folks about their farming practices is this deep bedding concept. Children that hold their noses and have to be forced to gather eggs suddenly want to go and sit in the coop for an hour at a time. Spouses battling over "those obnoxious chickens" suddenly fall in love all over again when deep bedding absorbs the odors by locking down the volatile nutrients.

Carbon is readily available. At worst, go to the landscape center and lug home a couple bags of peat moss. Get some bark mulch in a couple of garbage bags. My whole life I've been cultishly fiendish to find, acquire, and use carbon. It's paid off in 8.2 percent organic matter soils and rich pastures. In the city, you might not have pasture to put your finished product on, but I'll bet you could put it on some vegetables, sprinkle it around some trees, or side dress some hedge roots.

Once you leave the focus group and start down this practical path, you can truly become the solution we all desire. It won't be fantasy land any more. It won't be impossible. We're dreaming a possible dream here, a dream of practical world change that starts with one. That would be you. Will you join me?

Chapter 26

Regulations and Marketing

I'm not an attorney. That's the disclaimer.

Now on to capture San Juan Hill. We live in a time when government tyranny is exceeded only by citizen timidity. To complicate matters, as fears over everything escalate, we're fostering a climate of tattling, snitching and the like.

Our farm's first big run-in with the food police was when someone (probably a disgruntled neighbor) called the bureaucrats to report us for illegally selling meat. As bad as things are in the U.S. though, it's not nearly as bad as Canada.

I was up there speaking at a conference. After the speech, during a book signing session, a lady eased up to me, handed me a cloth shopping bad, and whispered in my ear: "Don't open it now; it's for later in your hotel room." I had no idea what it was, but thanked her and continued signing books.

When the session ended and my host was talking to the close-up and janitorial crew in the lobby of the auditorium, I opened the bag and snuck a peak: raw milk. And ice cold. Folks, nothing is like ice cold raw milk. It's ambrosia. My host and I exited and got in his car to go back to my hotel.

As we headed out of the parking lot, I told him about the delightful gift I had tucked between my feet on the floorboard. He jerked around and responded: "In Canada, it's illegal to transport raw milk on a public road." We both laughed for the next few minutes as we imagined me being incarcerated in Canada for transporting a quart of milk. The headline: "U.S. Lunatic Farmer Arrested and Charged with Illegally Transporting Raw Milk."

Perhaps the biggest disappointment in my country right now is how often folks ask "is it legal?" I don't know whether to be angry at the regulation bureaucracy, the society that clamored for that bureaucracy--oh, right, it's called oversight-- or the cowed people who think we need to ask permission to pee.

What happened to our rugged individualism? What happened to freedom, to independent agency? To civil disobedience, courage, and proceeding with what is right and noble regardless of its being blessed by the powers that be?

Whether it's backyard chickens, keeping chickens in your apartment in Manhattan, or butchering rabbits on the back porch, if it's the right thing to do it's the right thing to do. I'm a firm believer in the adage that it's often better to ask forgiveness than permission.

The thing to remember is that government regulators are not agents of "yes;" they are agents of "no." In my years of dealing with them, the one thing that seems consistent is that "no" is the default position. It makes sense. A regulator won't get a smack-down from superiors for saying no. Saying no means you don't have to figure anything out. It means you don't have to think through the situation. It means nobody will fuss at you for an alleged violation or something going wrong. If you don't okay the idea, you bear no liability for whatever happens.

Nothing can come back to bite you if you say no. Bureaucracy hates innovation. That's one of the thematic definitions of bureaucracy, which is all about keeping things the same. Don't rock the boat. Keep doing what we've always done. Change is death to paper pushers and bean counters.

I've been on a tear for many years to let farmers who want to process animals and sell them get an R2D2 that receives swab tests. Imagine a cute little Star Wars R2D2 sitting in your processing facility saying "swab test from batch number 4521 is okay." We can measure all sorts of pathogens and toxins with simple frequency tests. Great Britain is experimenting with infrared video with artificial intelligence in slaughterhouses to scan and catch pathogens without any bureaucrat on site.

When we faced down the chicken police who insisted that outdoor processing was "inherently unsafe," I challenged them to take samples of our chicken and compare them to what they were selling in the store. They weren't interested in empirical data. No, they wanted bathrooms, walls, toxic sanitizers. My question to them was "if it's clean it's clean. Who cares if it was wrapped in a million dollars of infrastructure or eviscerated in the kitchen sink?"

With the technology available today in instruments like spectrometers, we could easily institute empirical testing on food items. Just give me an R2D2 machine that eats my swab samples, sends the data to the state database, and if we exceed a benchmark come and see us. But to require massive amounts of concrete, stainless steel, hand-free washing stations, bathrooms, changing lockers and 24/7 temperature control without even being interested in empirical safety is asinine. When I approached a federal inspector with my R2D2 idea, he agreed immediately that it would work, but then caught himself: "oh, but then I'd be out of a job. No, I don't think I'd like that option." The awful truth is that the Food Safety Inspection Service (FSIS) is nothing but a jobs program.

What could be more consistent with "life, liberty, and the pursuit of happiness" than being able to raise my own food to feed my own body? Since when does anyone have the right to tell me I can't raise food on my own property? What could be more basic to personal liberty and being secure in our own effects than controlling through growth and processing what fuels our own being? That's the personal liberty angle. Some would consider it selfish thinking.

Okay, how about another angle? What could be more empowering than securing the means of production and processing of our food supply? Or asked another way, what could check the growing stranglehold of corporate food dominance better than homegrown fare? Folks concerned about food system centralization should do back flips when someone decides to grow their own food.

"Oh, but this isn't food. This is animals. It's not tomatoes; it's blood, guts, poop. Eeeeewww." My response: "Okay, let's get rid of all the pets; their waste is really foul. Let's exterminate the pigeons and wild birds. And do you know New York City now has volunteers patrolling the sidewalks with rat-hunting dogs to try to keep the rat population under control?"

In *Guns, Germs, and Steel*, bestselling author Jared Diamond documented the strength of civilizations and why some ascended and some descended. The germs part of the book explained how civilizations with proximate animal husbandry developed far more robust immune systems than those that didn't. Today, the medical professionals who have signed onto the hygiene hypothesis are certainly consistent with Diamond's argument.

The hygiene hypothesis is the idea that the immune system is like a huge muscle and without constant exercise it becomes lethargic. Exercise is another way of saying exposure to bugs. Richard Louvre picks up this same refrain in his wonderful book *Nature Deficit Disorder.* While he advocates the immunological strengthening of getting out in nature and not specifically commiserating with livestock, the similarities are obvious. Petting a chicken on a homestead is certainly equivalent to inhaling wilderness air carrying elk bugs.

Finland now leads the world in documenting the robust immune systems of children raised on farms versus the anemic immune systems of children raised in cities. Their research indicates that children literally touching and playing in manure in barns and toddling around on stable cleanings are far less prone to sickness than their cleaner city neighbors. We encourage folks to visit our farm not just for the aesthetics and sales, but to immerse in air and landscape speckled with bacteria and life. It's like an immune elixir fixer.

As a society, are we not in danger of succumbing to more sicknesses when we withdraw with an "Eeeeeewwww" from domestic livestock? New information developing from the microbiome, the actual diversity of bugs in the human gut, show an astounding connection with soil, intestinal bacteria, and both physical and emotional health. For good measure, I routinely drink out of the cow water trough, with the cows, to make sure I keep a healthy diversity of gut microbes.

I would suggest that people who want to stay an arm's length--or factory farm's length--away from domestic livestock actually invite sickness probability. Is it outrageous to pose the question, "is being disconnected from domestic livestock as detrimental to health as smoking or obesity?" Good data perhaps doesn't exist on this question, but it certainly is within the realm of possibility. Soap and water, not playing in human feces, refrigerators--these are all sanitation breakthroughs not to be denied. But that pendulum can swing too far, like anti-microbial soap proved.

Rather than being "Eeeeeewwww" to animals, maybe we need to be "Eeeewww" to folks who distance themselves from animals. Perhaps the greater egregious personal and societal error is in disconnecting rather than connecting. These are all points you may need as you graciously and diplomatically deal with friends, neighbors, and family members who push back against your new-found domestic livestock commitment. When you spread your wings and fly into a participatory and personally accountable food system, you need to be prepared for turbulence.

With all that said, to my knowledge no place in the U.S. prohibits processing your own animals for your own personal consumption, except perhaps Washington D.C. which may have such a prohibition. Nor does anything prohibit you from giving this food away. You can slaughter and process any animal for you and your family to enjoy without fear of any licensing. That's not to suggest you're free of prying eyes and finger pointing. But legally, fortunately, you have the right to process your own animals for yourself.

That processing does not have to happen on your property. If you live in the city in a condo and you need to process ten chickens and you don't have a good place to do it, you can take them to a friend's yard and do it. As long as the friend doesn't charge you money, you're okay. The law smiles favorably on charity and free good will favors, generally. If you live in an apartment owned by someone else who prohibits animals, you may be out of options. That's not a government issue; that's a private contract you've signed, which carries a lot more weight. You probably don't want to push that lease.

The problem comes when money trades hands. This is why these food safety and licensure laws ultimately aren't about food safety; they are about defining what's available in the marketplace. They're also about who gets to play the commerce game. The legal operative phrase is "in commerce." If you can figure out a way to make a trade or a bargain that doesn't neatly fit "in commerce," you've found a loophole.

The Farm to Consumer Legal Defense Fund (FTCLDF) and Institute for Justice are two organizations that litigate every day in this space to keep direct food trades possible. The FTCLDF maintains a 24/7/365 hotline for members to call in real time when harassed by regulators. Getting legal counsel in the crunch of the moment, when the bureaucrat stands there flashing a bronze badge, is an incredibly empowering thing. Everyone should join. I look forward to the day when these valiant organizations are as powerful as the National Rifle Association (NRA)--and that's not to denigrate the NRA. I'm a member. But when Americans care as much about food freedom as we do about gun freedom, we'll have a different food chain. In the big scheme of things, holding onto food is probably about as important as holding onto guns.

The only reason the authors of the Bill of Rights did not include food along with worship, speech, assembly, guns, and property was because they could not have envisioned a day when a homesteader couldn't sell a glass of milk or a homemade chicken pot pie to a neighbor. If you want to dig into this more, I suggest *Everything I Want To Do Is Illegal.* It's like drinking out of a libertarian fire hose; just sayin'. To be forewarned is to be fore-armed.

Even for a homestead, the right to food commerce is a huge issue because seldom can you make your production exactly match your home consumption. If you're savvy at all, you'll have overages. In the spring when all the chickens lay up the wazoo you'll be swimming in eggs. When the cow freshens and you have three gallons of milk a day--can your family drink all of that? You'll be sitting around like Templeton the rat in Charlotte's web, unable to move from tanking up on moo juice. We've already established that you want two pigs for their own social happiness; what if you don't need two hogs a year and only need one?

Even extremely small homesteads generally yield abundance beyond what you need for your domestic table. Though selling things may not have been at the top of your dream list when you first thought about homesteading, chances are you'll be wanting to trade something for money before long. What do the laws say about that?

That's where advice from the aforementioned legal non-profits can help. If I listed the legal minutiae of everything right now, it would be outdated before this book went to press. Instead, I'll give some general guidelines along with some loopholes.

Rabbits are such a small part of America's food system that they kind of slip through the cracks of licensure. If you want to process your rabbits at home, yourself (always my first suggestion) go ahead and sell them to anybody who wants to buy. Don't sell to a middle-man, like a grocery store. Make sure you sell it to the entity that's going to eat it, like a restaurant or an individual family. Unless you become a rabbit tycoon selling 5,000 a year, you'll probably never enter any government agent's radar.

Realize that most regulatory action is initiated on the basis of a complaint. Inspectors usually aren't wrecking balls looking for a fight. They take their kids to soccer practice and sing in the choir just like other people. While they often don't act like they have a heart, most do. You'll never get fined or imprisoned or arrested for violating any of these food safety regulations on your homestead. You might get a call; you might get a cease and desist order. You might get a visit. But the initial contact is investigatory and relatively courteous.

That said, you don't need to answer any questions. You can require a warrant before they visit. You can video all the interaction for a legal record. You can require all correspondence in writing--that'll buy you six months. You don't have to be mean, angry, or violent; you can force them to earn every step they take. If you're going to bow, make them at least sweat before you bow. As a matter of policy, on our farm we do not talk to regulators; we require everything in writing. Often that makes them go away. Too often regulators show up on your doorstep on a fishing expedition: "I heard from somebody who heard from somebody who heard from somebody who heard from somebody." On our farm, we nicely but firmly tell them they are trespassing and to get back in their car and leave. By all means don't sign anything. Don't give a statement. Bureaucrats love entrapment. Don't fall prey to their smiles and niceties; remember, they have guns.

Poultry enjoys a federal exemption for producer-grower processing and sales called PL90-492. It allows you to raise and process up to 20,000 head of poultry (includes, ducks, turkeys, chickens) and sell them without inspection to "end users." That's an important phrase. An end user is anyone who prepares and consumes the product on site. You can sell it to a deli, for example, but not out on the aisle in the meat case. If the buyer is the final stop for that product, like a restaurant, an institution, a hotel, an individual family, then that qualifies as an end user. Some states are more restrictive than the 20,000.

The only requirement in this wonderful exemption is that the poultry must be processed in a "sanitary and unadulterated" manner. Therein lies the problem, because what's sanitary to one bureaucrat may not be sanitary to another. This is what happened to us with our outdoor shed processing facility. The state inspector who gave us our initial green light retired and his successor disagreed and we suddenly found ourselves in a dogfight for survival.

Be assured that creativity is the name of the game when dealing with these rules. A sheet of plastic can be "an impervious wall." A wall can be any height. A window can be any dimension. A person can buy chicks and pay you a caretaker fee to raise them and you can process them for free. You can sell live chickens without any restrictions to anyone. As many as you want. The problem is processing; construct that portion of the deal so it's free.

Eggs are relatively unregulated unless you make claims. Even at our scale here at Polyface, we carry no licensing on our eggs because the carton makes no claim: grade A, size, etc. Just call them eggs and you stay free from inspection. People don't realize that inspection doesn't look at things people fear, like *salmonella*. It only looks at size, albumen viscosity, air sac size, and shell quality. None of that has anything to do with pathogens. But people bow to inspection like it's some sort of health protection. It's not.

Regulations make a huge distinction between poultry and meat. Poultry is not meat. Broilers are not meat. They're poultry. Meat is all the other domestic stuff: beef, pork, lamb, goat, llama, alpaca. Then you have wildlife: deer, elk, antelope, etc. Bison is kind of in a spot that straddles wildlife and meat. You can shoot a bison in the field and take it to an inspected abattoir and sell it under inspection, but you can't do that with beef, hog, lamb, or goat. If your head is spinning, you're in good company. These rules have no scientific backing whatsoever; they're just a hodgepodge of political expediency cobbled together since 1906 when Upton Sinclair wrote *The Jungle* and the meat packers begged Teddy Roosevelt to create the Food Safety Inspection Service to give them skirts to hide behind.

What no one wants to appreciate is that within six months of that book's publishing, the big eight meat packers lost 50 percent of their market. Half! Had Teddy Roosevelt never created the FSIS, this market adjustment would have altered the centralizing meat industry perhaps forever. Instead, the FSIS gave cover for these big outfits and the centralization and industrialization continued unabated. The industry/regulatory fraternity gradually tightened the noose on community abattoirs, canneries, and processors, aided and abetted by a society in love with convenience.

Where we are today is the offspring of convenience coupled with crony capitalism. If anyone or anything can break the power and corruption from this unholy alliance, it's thousands and thousands of lowly homesteaders processing animals in their own backyards and their own micro-facilities. If we're ever going to get our food system back, it will take thousands of new participants in love with animal production integrity and authentic food.

Wild animals carry no inspection requirements but you can't sell the meat. I can shoot a deer on a 70 degree F day, drag it a mile through the sticks, rocks, and squirrel dung, parade it prominently on the hood of my Jeep in the afternoon sun, butcher a week later in the backyard and feed it to my children and all their friends. That's perfectly safe and patriotic. But if I process a lamb or goat on an appropriate temperature day in the same yard with the same equipment, I can't sell an ounce to a friend at church. You tell me why that makes any sense.

On our farm, we harvest several deer every year and one of our favorite results of that is taking the meat up to an Amish outfit that turns it into tubes of summer sausage. It's wonderful; I think I could live on it indefinitely. But they can't do any beef; that's a regulated product. Why is a deer more safe than a beef? Again, this isn't about safety; it's about picking winners and losers at the marketplace table. Oh, and even though we have hundreds of customers who would love to buy it, we can't sell it. What a shame.

We've looked at "in commerce," "end user," "meat and poultry." Now let's examine "inspection." When you buy a package of ground beef at Kroger--oops, I should have said LOOK at a package of ground beef at Kroger, right?--it will have a little round circle that says "USDA Inspected" and will contain an establishment number. That's the inspected facility where the meat was killed.

With meat, three options exist. The lowest is called custom processing. This is a facility that processes meat for specific animal owners. In other words, I have a hog I want processed and I take the hog to this outfit; they process it and I come back and pick up the packages. Each package is stamped "Not For Sale." In other words, I can't sell that to a friend. I can give it away; I can eat it myself. But I can't sell it.

In this option, no inspector is on the kill floor sniffing and poking through the guts. Food safety liability is assumed by the owner of the animal, who will also consume it. From society's perspective, the owner is responsible for the quality of his own animal. Inspecting his own animal for him to consume would be like not allowing someone to eat a carrot out of her own garden without a bureaucrat saying it's okay. Don't laugh at that. I was in California speaking at a college and asked for a show of hands of how many thought the government should be required to inspect the vegetables from your own garden before you could consume them yourself, and a third of the hands went up. That's scary.

The loophole in custom processing is that I can sell an animal live, or even a quarter of an animal. As long as that live buyer's name, as the live animal owner, or portion thereof, is articulated to the custom butcher before the animal is slaughtered, that falls within the letter of the law as legal commerce. Technically, the owner is not buying packages; she brought in an animal, or portion thereof, and asked the custom processor to cut it into packages. No after-processing commerce occurs. That's legal because it was a live animal sale.

Next option: state inspection. Many states have given up their in-state programs but about half retain them still. Interestingly, the state requirements must be "equal to" the federal; the only difference is that meat cannot be sold across state lines. It can be delivered across state lines TO someone who already bought it or carried across BY someone who already bought it. But you can't take unsold meat across the state line and sell it. If it's not already sold, you can't carry it across state lines. You can't even possess unsold product in a delivery vehicle that crosses a state line on its way to re-entering the legal state where it will be sold.

The loophole here is to sell it on-line or on-location, making the sale happen in the inspecting state. The prohibition is only on sales across state lines, not transport across state lines. See how fun this is? You might ask why states even maintain this option if requirements must meet federal standards. One reason is pride. State bureaucrats don't like to give up territory, even to the federals. A second is that the industry likes this option because the state inspectors are generally much easier to work with than federal. The state bureaucrats at least have a state pride issue and want to make sure the state has market capacity. The federals have no such allegiances.

Under state inspection, a physical person must look at each animal slaughtered, examine the guts, and determine its wholesomeness for human consumption.

Next option: federal inspection. This is the ultimate enchilada. A personal inspector verifies the wholesomeness of each carcass. Before you assume this is done with any integrity, realize that in the large poultry plants, inspectors get less than one second to view the innards. You can't see a lot in one second. That's why routinely smaller abattoirs have a higher percentage of rejected carcasses than large processors. This condemnation percentage makes small plants look poor in the national data base. Consumer watchdog media admonish folks not to use small plants because they have more rejected carcasses.

Folks, those condemnations have nothing to do with overall ratios; they only reflect the fact that in small plants inspectors actually have time to see something. In the large ones they don't. This highly prejudicial government data disseminates into the populace, without the time distinction, and fuels the cries demanding "small plants need more oversight." You have to laugh to keep from crying.

With federal inspection, you'll get packages back with the little circle, called the blue buzz, and it allows you to sell it anywhere in the world. Now you understand the wild and wacky world of meat and poultry inspection and the general orthodoxy surrounding legal commerce.

With all those options, you'd think getting a beef or pig processed would be a piece of cake, right? You'd be wrong. As the cozy fraternity between government and industry fueled by consumer paranoia drove increasingly aggressive regulatory expansion through the 1960s and toward today, we lost thousands of community processing facilities. Today's renaissance of local food fueled by a new paranoia toward industrial meat and poultry is running far ahead of the surviving local capacity for processing. This is a crisis of national security.

Communities that can't feed themselves become subject to all sorts of manipulation by powerful interests. The first security is food security. Part of that is preserving the capacity to get animals into a cookable state. And by the way, the most dependency-predicated locally insecure food system is fake meat made in a techno-sophisticated multi-million dollar laboratory hundreds of miles away. Why environmentalists can't see the folly, just on a social, security, and logistical level, of this outrageous fake meat catastrophe is beyond me. Okay, Joel, settle down.

Plenty of home butchery books exist: buy a couple and launch. You can do lots of small amounts on the quiet. Go ahead. If you don't advertise, no one will know. When compliance becomes more difficult than circumvention, you know we've reached a new level of tyranny.

In my experience, anyone can sell the following amounts without really doing any marketing. If you just let your circle of acquaintances know what you have available, some will want to buy. I haven't done a scientific analysis to get these figures, but I've come to them as a result of the many homesteaders I've talked with over the years. In regulator conversation, social circles, work place chatter, anyone can sell annually, without much effort:

- 5 beef
- 10 hogs
- 4 lambs
- 2 goats
- 200 broilers
- 20 turkeys
- 30 dozen eggs a week

When you pass these numbers, you've become a farmer and are no longer a homesteader. And at that point, you'll need to start thinking more aggressively about legality, liability, publicity, risk, messaging, marketing and all the things that make business a joy ride.

I have no idea what will develop in the U.S. food system over the next ten years. I can tell you that most of the world does not live under the onerous food safety laws that we have here in the U.S. and other supposedly developed countries. If hard times come, your ability to funnel some integrity food into your neighborhood may be the difference between thriving and poverty. I know that in most places of the world, folks don't ask about a blue buzz and they're happy to get some sustenance.

For America, I don't hope for impoverishment, but I would like to see enough concern to free up the local stewardship-oriented authentic entrepreneurial food option. I'd love to see a FOOD EMANCIPATION PROCLAMATION to grant each person the freedom to procure the food of their choice from the source of their choice. And to see a rebirth of liberty-mindedness that realizes part of property rights is the right to sell it. If I can only sell a cow into proscribed channels, isn't that a taking, like eminent domain?

Be assured that all these issues will come front and center for you as you develop your livestock enterprises on your homestead. Be encouraged to seize individual sovereignty, to truly embrace market diversity, and to encourage personal choice. Injecting your stewardship and livestock passion into your community moves our collective needle toward resiliency, healing, and freedom more than you can imagine.

Thank you for indulging this exploration of what can be, of what should be, and indeed what ought to be a revival across our land. This is a time to move beyond dreams and implement the activities that will make our families stronger, our bodies healthier, and our land caressed. Together, with a million new homesteads and new in-condo chickens, we can move our foodscape and our landscape to a legacy of abundance and health for our children and grandchildren. May we live that they may look back, thank us for rescuing a tilting ecology, and from vibrant immunological function proclaim "my homesteading ancestors made a better life possible."

INDEX